T0245441

CAMBRIDGE LIBRARY COLLECTION

Books of enduring scholarly value

Life Sciences

Until the nineteenth century, the various subjects now known as the life sciences were regarded either as arcane studies which had little impact on ordinary daily life, or as a genteel hobby for the leisured classes. The increasing academic rigour and systematisation brought to the study of botany, zoology and other disciplines, and their adoption in university curricula, are reflected in the books reissued in this series.

Memoir of Sir Andrew Crombie Ramsay

Sir Andrew Crombie Ramsay (1814–91) was a British geologist with a particular interest in the effects of glaciation on the landscape. He travelled in Europe and America, and was a keen climber. His first work, *Geology of the Island of Arran* (1840), also published in this series, attracted the attention of Roderick Murchison, who found him employment with the Geological Survey, and Ramsay later succeeded Murchison as its director. He carried out important fieldwork in Wales, taught at University College London and the Royal School of Mines, and published a successful textbook. Another major contribution was his work on the origin of lakes: his controversial 1862 proposal that glaciers could hollow out lake basins even in the absence of earth movements was eventually accepted. Ramsay's younger colleague at the Geological Survey, Sir Archibald Geikie (1835–1924), who also wrote a biography of Murchison, published this memoir in 1895.

Cambridge University Press has long been a pioneer in the reissuing of out-of-print titles from its own backlist, producing digital reprints of books that are still sought after by scholars and students but could not be reprinted economically using traditional technology. The Cambridge Library Collection extends this activity to a wider range of books which are still of importance to researchers and professionals, either for the source material they contain, or as landmarks in the history of their academic discipline.

Drawing from the world-renowned collections in the Cambridge University Library, and guided by the advice of experts in each subject area, Cambridge University Press is using state-of-the-art scanning machines in its own Printing House to capture the content of each book selected for inclusion. The files are processed to give a consistently clear, crisp image, and the books finished to the high quality standard for which the Press is recognised around the world. The latest print-on-demand technology ensures that the books will remain available indefinitely, and that orders for single or multiple copies can quickly be supplied.

The Cambridge Library Collection will bring back to life books of enduring scholarly value (including out-of-copyright works originally issued by other publishers) across a wide range of disciplines in the humanities and social sciences and in science and technology.

Memoir of Sir Andrew Crombie Ramsay

ARCHIBALD GEIKIE

CAMBRIDGE UNIVERSITY PRESS

Cambridge, New York, Melbourne, Madrid, Cape Town,
Singapore, São Paolo, Delhi, Tokyo, Mexico City

Published in the United States of America by Cambridge University Press, New York

www.cambridge.org
Information on this title: www.cambridge.org/9781108037679

This edition first published 1895
This digitally printed version 2011

ISBN 978-1-108-03767-9 Paperback

MEMOIR

OF

SIR A. C. RAMSAY

D.Haine; Photo.

Walker & Boutall,Ph.Sc.

Memoir

OF

Sir Andrew Crombie Ramsay

BY

SIR ARCHIBALD GEIKIE, F.R.S.

DIRECTOR-GENERAL OF THE GEOLOGICAL SURVEY OF
GREAT BRITAIN AND IRELAND

WITH PORTRAITS

London

MACMILLAN AND CO.

AND NEW YORK

1895

PREFACE

THE life of a professional man of science seldom offers such variety of incident and interest as to justify more than a brief record. In most cases a summary of his work and an estimate of its value in the onward march of knowledge form for such a man the most fitting memorial. Now and then, however, a leader has appeared, who, by the fascination of his personality, or by the extent and importance of his individual achievements, has exercised so marked an influence on his contemporary fellow-workers, or on the general advancement of science, that the desire naturally arises to know something more of him and of his surroundings, than the mere list of his labours. One would fain learn how he came to be drawn into the ranks of the soldiers of science, and by what process of training or what stages of evolution he rose to be a captain in those ranks. The story of his discoveries may sometimes have had a vivid personal interest, and those who can best appreciate the value of these discoveries would gladly know how they were made.

The subject of the present memoir stood in the

forefront of the Geology of his time, and by the charm of his genial nature, as well as by the enthusiasm of his devotion to science, exercised a wide influence among his contemporaries.

To that large circle of friends who knew him in his prime, and to that yet wider public which recognises how much it has profited by his labours, some brief record of the life of Andrew Crombie Ramsay will be welcome. He was almost my earliest geological friend, and for many years we were bound together by the closest ties of scientific work and of unbroken friendship. It has been, therefore, a true labour of love to put together this little memorial of him. As far as the materials at my disposal would permit, I have allowed his personal experiences to be told in his own words. I have tried to trace the gradual progress of his development as a geologist, and to offer a short summary of what seem to me to have been the essential features of his contributions to his favourite science. And I have sought, though I fear with but imperfect success, to show something of that bright, sunny spirit which endeared him to all who came within its influence.

Sir Andrew Ramsay joined the Geological Survey when it was still in its infancy, and he remained on its staff during the whole of his active scientific career—a period of forty years. So entirely did he identify himself with the aims and work of the Survey, and so largely was he instrumental in their development, that the chronicle of his life is in great measure the record also of the progress of that branch of the public service. Recognising this intimate relation, I have woven into my narrative such additional detail as might perhaps serve to make the volume not only a personal biography, but an outline of the history of the Geological Survey of the United Kingdom.

Among those who have kindly supplied me with letters or information I would especially express my indebtedness to Lady Ramsay and Sir Andrew Ramsay's nephew, Professor William Ramsay, F.R.S., who have furnished many family and personal details; and to Mrs. Johnes and Lady Hills-Johnes of Dolaucothy, who have lent a large collection of letters. Old colleagues on the Geological Survey have likewise been helpful, especially Lord Playfair, Mr. W. T. Aveline, Mr. A. R. C. Selwyn, Professor T. M'Kenny Hughes, Professor A. H. Green, Mr. H. H. Howell, Mr. W. Whitaker, Mr. F. W. Rudler, Mr. A. Strahan, and the late Mr. W. Topley. Mr. M. J. Salter has lent a number of letters addressed to his father. To some of Sir Andrew's foreign correspondents I am likewise under obligation, particularly to Professor Zirkel, Professor Daubrée, Professor Rütimeyer, Professor Capellini, and the family of Signor Sella.

It has seemed to me that additional interest would be given to the biography by the insertion not only of a likeness of its subject, but of portraits of some of his more notable comrades. I have accordingly added likenesses of a dozen of his geological associates whose names and work are well known. These have been taken as far as possible from early photographs, so as to picture the men as they looked when they were actively engaged with Ramsay in geological work. But in some cases when no early likeness was available, or where the photographs had become too faded for reproduction, later portraits have been chosen.

GEOLOGICAL SURVEY OFFICE, JERMYN STREET,
LONDON, 12th September 1894.

CONTENTS

CHAPTER VII

CHAPTER VIII

CHAPTER IX

CHAPTER X

CHAPTER XI

LIST OF PORTRAITS

CHAPTER I

IN the little town of Haddington during last century several generations of Ramsays carried on the craft of dyers. At length one of the family, William by name, the son and grandson of previous Williams who had been content to pursue their calling by the banks of the East Lothian Tyne, determined to push his fortune in a wider sphere. He appears to have been a man of high principle and great energy, wide-minded and good-tempered, with a strong bent towards chemical pursuits, and not a little originality as an investigator. About the year 1785 he went to Glasgow, and became there junior partner in the firm of Arthur and Turnbull, manufacturers of wood-spirit and pyroligneous acid. Besides making dyers' chemicals and a variety of Prussian blue still known as 'Turnbull's Blue,' this firm was the first to manu-facture 'chloride of magnesia' as a bleaching liquor, and also 'bichrome.' Had William Ramsay patented some of his processes, it was generally believed among his friends that he might have become one of the richest men in the west of Scotland. But he did not consider himself entitled to retain for his own behoof a discovery which, if made widely known, would

B

benefit the general industry of the country, and he was content to remain comparatively poor.

The requirements of his business made him an excellent practical chemist, but his interest in chemistry reached far beyond these limits. In 1800 he founded the 'Chemical Society of Glasgow,' into which, by the energy of his example and the kindly courtesy of his manner, he brought those of his fellow-citizens who were interested in the progress of theoretical as well as practical chemistry. He was chosen first President, and among his associates were the well-remembered chemist and mineralogist, Thomas Thomson, Professor of Chemistry in the Glasgow University, and Walter Crum, of Thornliebank. Two years later, on the foundation of a wider brotherhood of science by the establishment of the 'Philosophical Society of Glasgow,' the Chemical Society was voluntarily dissolved in favour of the new organisation, which thus received, we may believe, not a little of the vigour which has enabled it to flourish till now as a centre of scientific life in the midst of the mercantile atmosphere of Glasgow. William Ramsay's reputation as a chemist spread outside his own country. His house was one of the attractions to foreign chemists who came to Glasgow; and even long after his death his widow received visits from such men as Liebig, who remembered her husband's meritorious work.

In the year 1809 William Ramsay married Elizabeth Crombie, a second cousin of his own, daughter of Mr. Andrew Crombie, writer in Edinburgh. The Crombies, like the Ramsays, had for many generations been connected with the trade of dyers. There is a tradition that during the famous Porteous Riot in Edinburgh in 1736, so graphically described in Scott's *Heart of Midlothian*, the mob, coming down the West Bow with their wretched victim, stopped at the shop of Crombie, the dyer, with the object of hanging Porteous from the pole above the door, when a shout arose that it would be a shame to do the deed at the door of so worthy a man. The crowd, determined as it was on vengeance, recognised the justice of this protest, and passed down into the Grassmarket, where they made use of the pole of another dyer not so popular among his townsmen. The last representative of the family who still carried on the trade of dyer in Edinburgh was a not less worthy citizen—John Crombie, who, firm in the ancient ways, went about in a tail-coat and 'stock' up to the end of his life, in 1874. He was a cousin of Sir Andrew C. Ramsay, who often stayed in his hospitable house during visits to Edinburgh.

Mrs. Ramsay was a woman of strongly-marked character, uniting a firmness of purpose with a gentleness and sweetness of nature that gave her remarkable influence over all who came in contact with her. Clever and wise, she had had her natural powers quickened and trained by an excellent education. She was beloved by the young, for whom her face used to light up with a cordial welcome. In the esteem and affections of her sons she ever held the foremost place. Her husband died in 1827, and her circumstances became thereafter somewhat straitened, but her cheery spirit and unruffled temper enabled her to keep a happy, though modest home for her children. She survived until the year 1858. The children of this marriage were four in number—Eliza, born in 1810, William in 1811, Andrew Crombie in 1814, and John in 1816.

In this well-ordered household, where both the father and mother had read widely, much was done to foster a love of literature among the younger members. It was one of the practices of the family that on at least one morning of the week French should be the language of the breakfast-table. On other mornings a paper from the *Spectator* would be read, or a passage from some standard English author. And doubtless the achievements of science, as far as they could be made intelligible and interesting, were often subjects of conversation.

Such was the household in which Andrew Crombie Ramsay was born on the 31st January 1814. Of his early years little record has been preserved. From his mother's letters we learn that when five years old, during a painful operation on one of his fingers, he showed such self-possession as to earn from the surgeon the encomium of being 'the most determined little fellow he had ever seen.' In a letter written to his wife in 1854, when his eldest daughter was a child, he says : ' I fancy I see Ella in the hayfield. These early days are never lost. I recollect them on rare occasions. I remember the first time I saw cowslips in a field ; how amazed and charmed I was ! The mind drinks in beauty in early life that never leaves it, if of good quality. Happy is the child whose first impressions are not of smoke, bricks, and gutters.'

For some time his health appears to have been delicate. At nine or ten years of age he was removed from Glasgow, and sent to the Parish School at Saltcoats, a little village on the coast of Ayrshire, where the sea-air might enable him to gain strength, and throw off his ailments. An observant boy could hardly have been placed in a position better fitted to stimulate his faculties. A sea-beach strewn with pebbles and shells lay in front of him, with rocks over which he could climb, and pools wherein he might bathe, or watch the movements of the creatures left by the tide. To the south a range of sand-dunes stretched for miles along the coast, mounting into ridges and sinking into hollows, which a young imagination could easily transfigure into ranges of mountains and lines of valley, interspersed with bare sandy plains and recesses that might typify trackless deserts—a lonely region, and a very paradise of boyhood. Then, in the interior, a long sweep of upland rose northward from the shore, commanding from its breezy heights a wide expanse of the Firth of Clyde, with the blue hills of Cantyre and Arran, sometimes even those of the north of Ireland, closing in the distance. On the lower grounds many a dell and ravine served as channels for streams which, haunted by trout and minnow, wandered through woodlands where many a bird built its nest, and where with the changing seasons came the successive attractions of blackthorn, mayblossom, blackberries, wild cherries, and hazel-nuts. There were likewise not a few ruined castles and crumbling peels, which an adventurous boy might climb, and where a contemplative one could find material for many a pleasant reverie. We can hardly doubt that surroundings such as these must have quickened in young Ramsay that love of nature, that delight in antiquities, and that devotion to out-of-door pursuits which formed such strong features in his character.

From Saltcoats he was eventually brought back to Glasgow, to continue his education at the Grammar School there. Mr. James King, probably his only surviving schoolmate, has kindly supplied the

following notes about his school-days: 'Andrew was
always cheerful and full of fun, so much so that he
was nicknamed "Appybe" (happy bee). He was
our leader in the stone-fights with the Camlachie
boys. He attended Mr. Dymock's class at the
Grammar School. When he was a child, a lady who
had called was telling Mrs. Ramsay what a good child
her lost son was, when Andrew, looking up to his
mother, said, "Mother, I would not like to be a good
bairn; good bairns aye die." He was very fond of
dogs. I remember his great grief at being obliged to
drown Puck for biting the postman.'

He lost his father in the summer of 1827. Twenty
years afterwards, on the anniversary of this sad event
in his life, he wrote as follows: 'My father died this
day twenty years at Roseneath. I was then between
thirteen and fourteen, and recollect it well. We had
been there about a week. He was very ill on the
way down in the steamboat, having had an additional
slight shock the very night before we started. Willie
was sent up from Roseneath a day or two before his
death. I accompanied him as far as Ardincaple
Ferry, and watched him across. It was a fine day,
but blew hard. On the way back I recollect playing
with flowers, so strange is it (I believe with all men)
that even in great distress the mind occupies itself
with trifles. I also recollect during this week of
severe illness my mother told me to take a book and
amuse myself. It was Shakespeare. I read *Julius
Cæsar*—the first play of Shakespeare I ever read, and
even then it highly interested me. Willie brought
down Drs. Coldstream and Buchanan with him. My
father died, I think, shortly after they arrived, having
been speechless for some time before. I did not see

him die, having, if I recollect right, left the room in
great distress some half-hour before.   My mother
prayed aloud soon after, most passionately and fervently ; so did Dr. Coldstream.   Curiously enough,
none of our relations came to aid the widow and her
children up to town, but Mr. Napier, the engineer,
came down of his own accord in one of his own
steamboats, and took on himself most kindly all the
arrangements.   My uncles arrived the day of the
funeral.   My mother threw herself into her brother
Andrew's arms, and said, " Oh Andrew !"

'The funeral was large and imposing.   He was
carried "shoulder-high" to the Ramshorn Churchyard, and buried in the Walkinshaw ground.

'By and by, shortly after, my troubles in life
began.   Willie was apprenticed to Napier, the
engineer, and I was sent to Mr. ——'s counting-house.'

The boy's education was thus prematurely cut
short, for in the straitened circumstances in which
the widow found herself after her husband's death,
she deemed it necessary that she should take boarders,
and that her sons should, as early as possible,
begin the active business of life.   Andrew was
intended for a mercantile career, and went when a
mere boy into the office to which he refers in the preceding extract.   After being some time there he
removed to the warehouse of a firm of linen merchants
in Glasgow—a situation in which he seems to have
been specially unhappy; for mention of the misery he
there endured occurs in his diaries and in his family
correspondence long years after he had become a
successful man of science.   He once came upon one
of these old masters of his in a little inn in Wales, and
the following entry occurs in his journal of that day :

'After dinner an old man, whom I had observed promenading the road before the inn, came into the room and took off his hat; his hair was bleached. In an instant a recollection flashed upon me. I started up and stretched out my hand, crying, " Mr. ——, I am delighted to see you," for my heart warmed towards him, in spite of all his want of consideration and kindness when long ago I sat, a boy, at a desk in his office. How changed care and anxiety have made him! He is an old, old man, though only sixty-one, and has been very ill.'

There never appears to have been any question in the family but that Andrew was to devote himself to mercantile pursuits. Yet, from the very outset, he kept his interests broad, and made amends for his curtailed education by cultivating his mind with wide reading. His natural tastes led him to continue the literary pursuits that had from his early years been so well fostered at home. He was an omnivorous reader, and acquired a facility in expressing himself in clear, vigorous language.

An interesting relic of this period of his life has survived in the shape of a few numbers of a manuscript periodical, written by him and a few young men of similar tastes. He acted as editor, and the paper circulated among the families and friends of the contributors during the years 1835 and 1836. It bore the name of ' Ramsay's Miscellaneous Journal,' and upon the wrapper of each number, in the handwriting of the editor, some appropriate motto appeared from a play of Shakespeare or a poem of Pope. The articles contributed by him included some nightmare hallucinations and sketches of character, with occasional sonnets and odes, more or less grotesque in subject and treatment. The concluding number closes with an editorial farewell: 'May our journal rest quietly in its grave; and if ever its pages should be used to light your pipes, peace be with its ashes!'

Though he had not himself matriculated at the University of Glasgow, he came into close personal relations with some of its professors and many of its students. Chief among his academical friends and advisers was Dr. J. P. Nichol, the well-known and accomplished Professor of Practical Astronomy. To this sympathetic associate he owed more than to any other for the guidance and encouragement which eventually led him into the career of a man of science. Among the young men then attending the University his closest friend was Lyon Playfair, now Lord Playfair, who was one of the boarders in Mrs. Ramsay's house.

In pursuance of the intention that he should follow a mercantile profession, Ramsay, about the year 1837, entered into partnership with a Mr. Anderson as dealers in cloth and calico. The firm took an office in the Candleriggs of Glasgow, and carried on business for some three years. But the venture was not successful, and the copartnery was dissolved, leaving Ramsay poorer in purse, somewhat enfeebled in health, and rather depressed in spirits.

It was natural that these successive disappointments should create a strong revulsion in his mind against an occupation which had never had great attraction for him. In a letter to his brother William, written in 1846, when he had thoroughly established his position in the Geological Survey, he refers to these early and bitter experiences of his life: 'You

must bear in mind how unhappily I was placed —
first with ——, when a system of miserable petty
tyranny was carried on from beginning to end, with
other disagreeables going much against the grain;
then with ——, a falling, low concern from the
beginning, and then something still worse behind.'

The island of Arran has been for the last two or
three generations one of the chief centres of attraction in the west of Scotland. To the inhabitants of
Glasgow it has offered a much-prized retreat, where
pure air and charming scenery can be reached after a
journey of only a few hours. It was the custom of the
Ramsay family, and of many families of their acquaintance, to spend as much of the summer as possible
in this delightful island. In those days the accommodation to be had in Arran was of a far more primitive
kind than it is generally now. Almost the only
available lodging was to be found in the little thatched
cots of the peasantry, and the unpretending farm-
houses, where the rooms were few and small, and the
furnishing generally scanty. Yet into one of these
lowly dwellings a large family would contrive to
squeeze itself, laughing at the discomfort with the
light-heartedness of holiday-makers who were prepared to enjoy everything. The conventionalities of
town life were left behind. Except for the hours of
meals and of sleep, and the intervals of bad weather,
the time of the visitors was spent entirely out of
doors. Bathing, boating, climbing, and walking or
driving to different parts of the island filled up each
day, and the evenings brought pleasant interchanges
of hospitality, with music and dance and endless merriment. If at the end of the week the heads of families
brought down with them more guests than the capa-

cities of the cottages—elastic as these were—could accommodate, there was always the homely and comfortable hostelry of Mrs. Jameson to fall back upon, with the calm bay in front, the Castle woods behind, and the noble cone of Goatfell towering into the sky beyond them.[1]

Sixty years have passed away since the time to which I am now referring; and though in this interval Arran has altered far less than other places on the Firth of Clyde, it has, nevertheless, undergone some marked changes. The old village of Brodick, for instance, with its long row of thatched cottages, has been removed. The old inn no longer 'invites each passing traveller that can pay,' though the build-

---

[1] Among the reminiscences of this pleasant Highland inn I recall the eccentricities of a half-witted but pawky attendant, who used to be employed in miscellaneous errands, and had a specially pronounced love of brandy. On one occasion he was pushing his boat down the beach, when two visitors came up and asked where he was bound for. He answered that he was going across the bay to the Corriegills shore for a bag or two of potatoes. The gentlemen asked to be allowed to accompany him ; a request with which Sandy willingly complied, the more especially as they volunteered to do the rowing if he would steer. Having crossed the bay, they were coasting quietly past the huge boulder of granite which, lying on the red sandstones, forms so notable a landmark on that part of the shore. Directing the attention of his crew to this object, Sandy remarked : 'Maybe ye'll no believe me, but if anybody climbs to the tap o' that stane and cries as loud as he likes, there's naebody can hear him.' This statement, as he expected, was received with a smile of derision, whereupon he insisted that he would wager them a bottle of brandy that it was true. So they drew to land, and Sandy, jumping ashore, was speedily on the top of the boulder, where he proceeded to open his mouth and swing his body as if he were roaring with the strength of ten bulls of Bashan, but without emitting a sound. 'Very extraordinary,' said his friends, and they resolved to try the experiment themselves. So when Sandy had descended, they proceeded, with rather less agility, to clamber up the stone. When they were both on the top they proceeded to shout with such vehemence that they might have been heard on the other side of the bay. Sandy, however, stood on the shore below, putting his hand behind each ear in turn to catch any sound that might come from the boulder. They shouted to him until they were nearly hoarse, without evoking one sign of recognition from him. At last coming down they demanded if he meant to say that he had never heard them. Sandy had a remarkable power of expressing astonishment by his mere looks, and availing himself of this power, he loudly protested that he had never heard one single sound from them, and, with a face of childlike innocence, asked if they really had called out. He was allowed to pull the boat back himself, but he had his bottle of brandy that evening.

ing still stands as part of the offices of the Castle.
The deserted pump-well remains to mark the centre
of the life of the vanished hamlet. A large hotel,
with waiters and other products of modern civilisation, has since risen at Invercloy, on the south
side of the bay, together with many slated houses;
while the inns all over the island, as well as the
farm-houses and cottages, have been much enlarged
and improved. The young visitors of to-day would
probably look with disdain on the humble cots where
their mothers and grandmothers were contented and
happy. But it may be doubted whether the charms
of this most delightful of islands are more appreciated
than they were in old days when the enjoyment of
them was coupled with discomforts now happily
removed.

Since the early decades of this century Arran has
enjoyed a special reputation as a field for geological
study. Its mountainous northern half has been held
to represent the main structural features of the Scottish
Highlands, while its southern half has been regarded
as affording examples of the younger formations, and
especially of the igneous rocks, which form a con-
spicuous feature in the geology as well as the scenery
of the southern part of the opposite mainland. It has
been described as affording an epitome of the geology
of Scotland, with all the salient points of structure
comprised within such narrow compass, and so clearly
displayed as to afford exceptional facilities for practical
investigation. Its coast-line supplies an almost continuous section of the rocks, with admirable exposures
of their various structures and relations to each other.
Its streams, too, coursing for ages from the watershed
to the sea, have trenched their channels into the solid

rock. All over the island, crags and rugged knolls
reveal the nature of what lies beneath the surface,
while the peaks and crests of the northern mountain
group form the background of the finest landscapes.
Nowhere can the influence of geological structure
upon scenery be more clearly seen, and nowhere is
that influence displayed in forms that more em-
phatically appeal to the imagination. It is a region
where a slumbering love of geological inquiry can
hardly fail to be stimulated into activity, and where
a latent aptitude for such inquiry may easily be
quickened into life.

Such were the surroundings amid which Ramsay
spent the holidays of his boyhood and youth. I have
not been able to trace definitely the beginning and
earliest development of his enthusiasm for geology.
There can be little doubt, however, that, over and
above the effect of his environment, he owed much of
the impulse which led him into the geological field to
the influence of two early friends. When still a boy
at Saltcoats, he had come into close contact with
David Landsborough, with whom he then began a
life-long friendship. This genial man and enthusiastic
naturalist, born in 1779, became in 1811 minister of
the parish of Stevenston, in which part of the village
of Saltcoats lies. He had from an early period of his
life devoted himself to the study of the botany and
natural history, not only of his own parish, but of the
neighbouring region of Ayrshire and of Arran. So
ardent was his devotion to these pursuits, and so
successful his cultivation of them, that he was known
as the Gilbert White of the west of Scotland. He is
said to have added nearly seventy species to the previously known flora and fauna of Scotland. His

personal influence in communicating the contagion of
his love of nature is vividly remembered by those who
knew him.   As Ramsay came under this influence
when a mere boy, we can hardly doubt that it helped
in giving the bent to his future life-work.

The other friend, who contributed still more to
the determination of Ramsay's geological career,
was Professor Nichol, already referred to.   Besides
guiding the young man's reading, this helpful mentor
incited him to the undertaking of definite pieces of
geological field-work.   Nichol, though not a professed
geologist, had himself read widely and critically in
geological literature ; he was therefore well qualified
to suggest lines of inquiry, to appreciate the significance of new observations, and to share in the plea-
sures and excitements of geological rambles.   He, too,
used to spend his summer holidays in Arran, and while
there enjoyed long walks and talks with his young
friend.   If any stimulus to sustained geological effort
had been needed on Ramsay's part, it was amply
supplied by 'the Professor.'   When the two friends
were separated, long letters of suggestion and advice
would come from Nichol.   The kindly and helpful
interest thus taken in him was always gratefully
remembered by Ramsay, who never ceased to look
back upon the Professor of Astronomy as his true
father in science, to whose wise counsel and assistance
he owed the happy change from a merchant's office to
the life of a professional man of science.

The fame of Arran as a happy hunting-ground
for the geologist drew many men of note to visit
it.   Of one of these visits Lord Playfair has
been so good as to communicate the following recollections :—

At the latter end of April 1836, or beginning of May in that year, I was going down to Arran, and was reading Lyell's *Geology*, which I had got as a prize at Graham's Class of Chemistry. Sitting beside me in the steamboat was a charming lady, who entered into conversation with me, and I showed her my book. I expressed great admiration for the author, and she smiled, and then called a gentleman from the other side of the steamer, to whom she introduced his young admirer. This was my first introduction to the Lyells. At Arran I used to help Mrs. Lyell in collecting shells, for at that time I knew something of conchology, while Lyell geologised in the interior of the island. Ramsay joined me in Arran after a few days, and I told Mr. Lyell that my friend would like to help him in his excursions, which thereafter they used to make together.

In the letter to Lyell, given at p. 92 of this Memoir, Ramsay himself dates the beginning of his serious study of geology from about the year 1836, and acknowledges his great indebtedness to the illustrious author of the *Principles of Geology*.

It is not possible now to recover traces of the successive tours and excursions by which the young geologist gradually filled up the geological map of Arran. He had been preceded by several able observers, who had published accounts of the structure of the island, notably by Macculloch, Jameson, Sedgwick, Murchison, and Necker de Saussure. But their descriptions could not be regarded as more than outlines of a wide subject, which would require years of patient research before its details could be mastered. It was with no idea of testing, still less of criticising, their labours that Ramsay followed in their footsteps along the shores and up the glens. He had not originally proposed to himself to publish any of his observations, which were made entirely for the pleasure they brought in their train, as they led him year after year over hill and dale. Gradually he found that various facts met with by him in the course of his rambles had

not been noticed by others before him. Thus, as far
back as the summer of 1837, he had observed the
mass of granite of 'Ploverfield,' of which the first
published account was given three years later by
Necker de Saussure.[1] These discoveries were duly
communicated to his friend Nichol, who doubtless
made good use of them as an encouragement to
continued investigation.

The meeting of the British Association for the
Advancement of Science was held in Glasgow in
September 1840. Among the preparations for that
meeting a committee was started for the purpose of
gathering together a collection of specimens, maps,
and sections illustrative of the geology of the west of
Scotland. In order to expedite the task, various
sub-committees were formed, to each of which a special
branch of the work was assigned. One of these was
organised to prepare a model of the island of Arran,
together with specimens of its geological formations.
The convener of this sub-committee was Professor
Nichol, who, as one of the local secretaries of the
Association, undertook a large amount of labour and
responsibility, and contributed much to the success of
the Glasgow meeting. Ramsay was the secretary of
the sub-committee, and, single-handed, did almost the
whole of its work. In reporting to the general
Museum Committee what they had done, Professor
Nichol, who drew up the statement, remarked that
'Arran had previously been surveyed by several
geologists ; but although these eminent men had given
valuable accounts of their observations, many blanks
remained to be filled up, and several important questions, having reference to the particular modes and

[1] *Trans. Roy. Soc. Edin.* xiv. (1840), p. 667.

epochs of the elevatory movements, and other phenomena of which this remarkable island is a memorial, do not appear to have been stirred at all. The Committee cannot presume that all deficiencies are now supplied, but they are certain that many points formerly obscure have been illustrated by their labours, and that a foundation at least is laid for a very complete and singular geological monograph. Their specimens, amounting probably to 700 or 800, have been selected with much care,[1] many sections have been drawn, a large map is in progress, and they have every hope that the model will, when finished, answer the purpose of rendering a great class of phenomena more palpable than could be done by any other mode of representation. It is necessary to mention that a new survey of the island in every locality has been executed, and that nearly all these labours have been gratuitously performed by their secretary, Mr. Andrew Ramsay, to whose talent and untiring energy their success is wholly owing.'

By the time the Association met, these active preparations had been completed. The specimens from Arran, after much anxious consultation over them on the part of Professor Nichol and his young associate, were duly displayed, the large map and sections were hung up, and the model, on the scale of two inches to a mile, was exhibited, with all the geological formations of the island clearly depicted on it in distinct colours.

A notable gathering of geologists assembled in Glasgow in September 1840. They included Lyell,[2]

---

[1] These specimens became the property of the British Association, and were handed over to the Andersonian University of Glasgow. But many years afterwards (1876) the Andersonian authorities, having no longer room for them, returned them, and they are now in the British Museum.

[2] Charles Lyell, born 1797, died 1875; author of the immortal *Principles of Geology*.

C

Greenough,[1] Buckland,[2] Phillips,[3] Murchison,[4] De la
Beche,[5] Smith of Jordanhill,[6] Agassiz,[7] Strickland,[8]
Edward Forbes,[9] and Griffith.[10]   It was before this
audience that Ramsay read his first scientific paper,
' Notes taken during the Surveys for the Construction
of the Geological Model, Maps, and Sections of the
Island of Arran.'[11]   In this communication he gave a
brief sketch of his work.   How he was guided in the
conception of it, and in the further elaboration of his
results, will appear from the following sentences in a
letter to him from Nichol: ' In writing out your
memoir never omit to draw attention as you go along
to points yet requiring elucidation, and which present
hopes of something very interesting.   You must make
this memoir short, chiefly in the way of scientific notes.

[1] George Bellas Greenough, born 1778, died 1855; one of the founders
and the first President of the Geological Society of London.
[2] William Buckland, born 1784, died 1856; author of *Reliquiæ Diluvianæ*,
also of one of the most celebrated Bridgewater Treatises, and of numerous geological
memoirs; Dean of Westminster, and Reader in Geology in the University of
Oxford.
[3] John Phillips, born 1800, died 1874; one of the founders and for many
years General Secretary of the British Association; was for some years attached
to the Geological Survey under De la Beche; an able writer and clear lecturer
on geology; succeeded Dr. Buckland in the geological readership at Oxford.
[4] Roderick Impey Murchison, born 1792, died 1871; author of *The Silurian
System*, etc., and Director-General of the Geological Survey from 1855 to 1871.
[5] Henry Thomas De la Beche, born 1796, died 1855; founder and first
Director-General of the Geological Survey of Great Britain; author of some
valuable papers and treatises.
[6] James Smith, born 1782, died 1867; author of a remarkable work on *The
Voyage and Shipwreck of St. Paul*, and some of the earliest papers on the shelly
deposits of the Glacial Drift.
[7] Louis Agassiz, born 1807, died 1873; famous as a writer on fossil fishes,
and for his contributions to glacial geology.
[8] Hugh Edwin Strickland, born 1811, died 1853; a geologist of great
ability and promise; killed by a passenger train when examining a cutting on the
Manchester, Sheffield, and Lincolnshire Railway.
[9] Edward Forbes, born 1815, died 1854; one of the foremost British
naturalists of his time; attached to the Geological Survey, and shortly before
his early death appointed Professor of Natural History in the University of
Edinburgh.
[10] Richard Griffith, born 1784, died 1878; the most illustrious geologist
Ireland has produced.
[11] *Report*, Brit. Assoc. 1840, Sections, p. 92.

The popular descriptive part must be kept for your book. I have been thinking of a speculation with Griffin in regard of your model and book, which I think might be very advantageous to you; of which, when we meet.' We shall see immediately what came of the 'speculation' here referred to.

Ramsay was heartily welcomed at Glasgow into the brotherhood of geologists. He formed there some of the most lasting and influential friendships of his life. It was there that he first came in contact with De la Beche, under whom, though at that time undreamt of by either of them, he was within a few months to enter upon the career of a professional geologist. It was there that he first met Murchison, who was from the very first deeply impressed with his capacity and geological ardour. It was there, too, that he made acquaintance with Edward Forbes, and began that intimacy which linked the two men together by the closest ties of friendship in the prosecution of scientific work.

It was arranged that on the Saturday of the Association week an excursion should be made to Arran, and the young geologist who had explored the island so well was invited by Lyell, who was President of Section C (Geology), to read his communication the day before, in order that those who intended to take part in the excursion might be put in possession of the necessary information. The excursionists divided themselves into two parties, one proceeding direct by steamer, the other by railway to Ardrossan, and thence by steamer to Arran. Ramsay was to conduct the party united at Brodick, and give them a general exposition of Arran geology. But he had worked hard in making all the preliminary preparations, and

for some days before had been up early and late.
Hence, when the morning came, he unluckily overslept
himself, and was too late for both steamboat and train.
This untoward accident he never ceased to regret.

The excitement of the Glasgow meeting, the first
entry into the company of renowned geologists with
whose names he had so long been familiar, the first
public exhibition of his own work as a geologist, the
first plunge into the sea of active scientific discussion,
and the cordial welcome extended to him by men
whose achievements he had followed from afar, left
Ramsay with many regrets when he came back again
to the consideration of his own prospects in life. How
gladly would he have taken to science as a calling if
only any opening had offered itself. Nine years afterwards, when he had found his place in the active
brotherhood of men of science, chancing to meet
Professor Johnston of Durham, whose acquaintance
he had made at the British Association in Glasgow,
he recalled to him an incident of that meeting,
which he thus describes : ' On the Sunday of the
Association week I chanced to overtake Johnston in
Ingram Street, and, talking about geological matters,
I told him how I was busy with mercantile affairs, and
longed for an opportunity to engage in geological
pursuits, after the happy taste I had had of it in
working before the coming of the Association. " Stick
to your work," quoth he, "and don't forget your
geology, and something may arise ! " He spoke truly.'

The British Association meeting, while it had
stimulated his bent towards geological work, threw no
light upon the dark outlook before the young man.
From a letter of his mother's, it appears that there was
at one time some prospect of his going out to Tasmania,

and with her maternal desire to keep all the family around her if that might be, she gladly welcomed any proposal that would prevent such a breaking up of her home-circle. As one disappointment succeeded another in his efforts to obtain a solid footing in business, he employed himself in completing his account of Arran. The 'speculation' referred to by Nichol took formal shape in an agreement between the Glasgow publishing firm of Richard Griffin and Co., and Andrew Ramsay, ' Merchant in Glasgow,' dated 2nd November 1840, by which, in consideration of the payment of a sum of twenty-one pounds, the latter undertook to prepare within three months a work on the geology of the island of Arran, together with the necessary views, sections, and maps.

In pursuance of this agreement, the work was duly written, and appeared the following spring as a thin octavo volume of seventy-eight pages, with a little map, a page of sections, and upwards of two dozen woodcuts, chiefly from drawings by the author. It was entitled *The Geology of the Island of Arran from Original Survey, by Andrew Crombie Ramsay*, and was appropriately dedicated to Nichol. This essay has long since taken its place among the classics of Scottish geology. As a broad outline of the structure of an exceedingly interesting geological region it was a most meritorious production. It gave sufficient detail to show how carefully its author had gone over the ground, how accurate and acute he was as an observer, and how clearly he saw the relation between scattered or isolated facts and the broad principles that connected them. While his chapters did not by any means exhaust Arran, they correctly described its general geological structure. In particular, the existence of

a 'New Red Sandstone' series, first proposed by
Sedgwick and Murchison, was clearly recognised by
him. Other observers have since disputed this
assertion, but its truth has recently been confirmed
by the Geological Survey. The history of the igneous
phenomena remains very much as Ramsay left it, and
is not likely to be much advanced until the still comparatively unknown southern part of the island is
mapped in minute detail.

Apart from the excellence of his essay as a geological
treatise, it had no little merit as a piece of descriptive
prose. A few passages from it may be quoted here
to show that, besides cultivating habits of geological
observation, the author entered thoroughly into the
spirit of the scenery amid which he was working, and
could depict in graphic words the aspects of the landscapes. Let us accompany him to the top of Goatfell,
the highest summit in Arran, and listen to his account
of it : 'The eye of the geologist suddenly rests on a
scene which, if he be a true lover of nature, cannot
fail to inspire him with astonishment and delight.
The jagged and spiry peaks of the surrounding mountains ; the dark hollows and deep shady corries, into
which the rays of the sun scarce ever penetrate ; the
open swelling hills beyond, the winding shores of Loch
Fyne, and the broad Firth of Clyde, studded with its
peaceful and fertile islands ; the rugged mountains of
Argyllshire, and the gentle curves of the hills of the
Western Isles, their outlines softened in the distance,
form a scene of surpassing grandeur and loveliness.
In all its varying aspects, it is a scene, the memory of
which can be dwelt on with pleasure : whether it
be seen in the early morning, when the white mists,
drawn upward from the glens, float along the hills, and

half conceal their giant peaks; or in the gloom of an
autumn evening, when the descending clouds, urged
onwards by the blast, flit swiftly across the mountain
sides, while ever and anon their gloomy shoulders
loom largely through the rolling masses, and seem to
the beholder to double their vast proportions; or in
the mellow light of a summer sunset, when the shadows
of the hills fall far athwart the landscape, and the
distant Atlantic gleams brightly in the slanting rays of
the setting sun; while, as he sinks below the horizon,
it is difficult to distinguish the lofty summits of Jura
and the Isles from the gorgeous masses of clouds
among which he disappears.'

In the midst of these impressive scenes, while
enjoying to the full their picturesque beauty, Ramsay's
eye was ever keenly sensitive to the geological lessons
so vividly taught by them. Lingering among the
granite precipices, and 'surrounded by the grey peaks
of the solemn hills,' the observer reflects that these
colossal features in the scenery, notwithstanding 'all
their appearance of majesty and power, are day by day
slowly crumbling into dust. Even now the landscape
on which he mutely gazes is imperceptibly yielding to
the never-dying principle of change; and the time will
come when, with all its varied features, it shall have
passed away, and left no trace behind.'

The young geologist had an eye, too, for the little
touchès of human pathos which so often lighten up the
sombreness of a Highland scene. As he comes down
North Glen Sannox, once a populous valley, but in
his day, as it is still, almost uninhabited, he contrasts
its very different conditions. He marks how 'green
spots, clothed with a close-cropped herbage, and still
bearing witness to the marks of the plough, surround

each ruined clachan. The hazel and the fragrant birch, the ash and the charmed rowan, fringe the banks of the stream, or mark the remains of the little garden-enclosures; and mingled with these may be seen the white blossoms of the gnarled elder, famed of old for its irresistible power in scaring the midnight witches from the neighbourhood of lonely dwellings, and counteracting the malicious pranks of the fairies, who, it is well known, still inhabit these desert wastes!'

The author avoids letting his own personality be seen in the course of his narrative, but in the following passage we seem to meet him coming back somewhat jaded from a long tramp to his welcome resting-place for the night in the snug homely inn of Loch Ranza. 'Tired and hungry though the traveller be, and with the very smoke of the little inn curling before his eyes, let him pause for a moment at the entrance of the loch, and seating himself on a granitic boulder, quietly contemplate the placid scene before him. Trees there are few to boast of, and what is pleasanter, there are still fewer strangers, for to the traveller in such a scene, all strangers seem out of place but himself. The sinking sun shines bright on the gleaming peaks of Caistael Abhael and Ceum na Cailleach, where the shadows of the rugged scars and deep hollows of the winter torrents, mingled with the lights brightly reflected from the projecting rocks, form a hazy radiance which more obscures than illuminates the shady recesses of the rugged corries. The tide is at its full, and the lazy sails of many a lagging fishing-boat, the image of the ruined tower and of the green hills around, lie calmly reflected in the unruffled waters :—

The lake returned in chastened gleam
The purple cloud, the golden beam;
Reflected in the crystal pool,
Headland and bank lay fair and cool;
The weather-tinted rock and tower,
Each drooping tree, each fairy flower,
So true, so soft, the mirror gave,
As if there lay beneath the wave,
Secure from trouble, toil, and care,
A world than earthly world more fair.

'But it is in a cold February evening that the pleasant solitude of the place will be most esteemed. There, seated at a blazing peat-fire, as the geologist extends his notes or arranges his specimens after his day's work, he will hear the piercing wind whistling down Glen Chalmadael and the narrow pass of Glen Eisnabearradh, then dying away as it reaches a wider expanse of the loch, to be again renewed by a louder and a shriller blast. And as he loiters to the door to speculate on the probabilities of the morrow's weather, he may chance to see the burning heath, like the beacons of old, blazing on the hills around, and faintly gleaming on the far-distant headlands of Argyllshire.'[1]

It was while Ramsay was engaged in the preparation of these chapters for the printer that the long-looked-for prospect of congenial employment at last opened out to him, in a form as unexpected as it was welcome. Among those who, from what they had seen of him and his work at the British Association, had formed a high opinion of his geological capacity was Murchison. This illustrious geologist, then in the full tide of his work among the older formations of the north and east of Europe, had entertained the idea of possibly extending his labours

[1] *Geology of the Island of Arran*, pp. 7, 27, 36, 40.

to North America, though he ultimately went to Russia instead. The young geologist who had done such excellent work in Arran would, he thought, make an admirable companion and assistant in his foreign expeditions; and in the autumn he wrote to propose such an employment to his young friend. No letter appears to have survived from Ramsay himself in reference to this sudden lifting of the clouds that had darkened his path. But we get a glimpse into the family circle in a letter written at the time by his mother to his brother William. 'Dr. Nichol,' she says, 'seems to think Andrew will have to go to London about the beginning of February. Andrew is in high spirits himself with the prospect. I hope it may turn out as much for his good as he expects. For my own part, I think there should be some written agreement about money matters; it is far more agreeable to claim as a right than to get as a favour, although the very travelling at Murchison's expense is a matter of consequence, and you may say although he were to get nothing he will see the world. At the same time, as he cannot afford to be without a salary, I hope it will be given.'

After some delay all the preparations were made, and Ramsay left home for his new career on Monday, 15th March 1841. A large band of his old friends and associates assembled on the Broomielaw to see him start, for he had arranged to take steamboat to Liverpool, and pay a visit there on his way to London. His journey and subsequent doings are best told in his own words :—

LIVERPOOL, WEDNESDAY (17*th March* 1841).

MY DEAREST MOTHER — You have by this time got over the first violence of your sorrow at parting with me, and however painful the separation is to all of us, you will find that time will gradually accustom you to my temporary absence; and you will look on a letter from me in the same light as you do one of Johnie's, with this difference, that you have the absolute certainty of seeing the second son (the go-between—the link between Willie and Johnie—who has part of the features and part of the character of both) in less than a year, and probably in six or eight months. Won't I rush home to Glasgow? brimful of London and Russia—of sights, wonders, and travels, a perfect Munchausen, telling most incredible stories about bearded Muscovites, horrible escapes from bears and wolves, burning suns and mountains of snow, expatriated Poles and Siberian mines. How I was introduced to the Emperor, how he smiled and bowed, and by a smile and a bow secured a deathless immortality, and honourable mention in the 2 vols. royal 8vo which are to hand down to future times the results of Mr. Hosie's[1] observations on men and manners in Russia; for a bow from a prince to a geologist excuses the depopulation of Poland, and a smile renders him amiable and attractive in the bosom of his family and in all his private capacities.

LONDON, 25th March 1841.

MY DEAR WILLIE—You will have heard all about

---

[1] 'Andrew Hosie' was a nickname by which he was familiarly known among his friends and associates. In another manuscript journal named 'The Renfield Rocket,' of later date than the 'Miscellaneous Journal' already referred to, the scientific doings of this personage are made the subject of jocular description. A Scots song also appears there to celebrate his virtues, of which the refrain runs—

My Hosie O! my Hosie O!
He's neither thin nor brosy O!
There's no a lad in Scotland broad
Can ever match wi' Hosie O!

me ere this from our folks at home, but perhaps I
may as well give you a synopsis of the whole of my
proceedings. I left Glasgow on Monday, and arrived
in Liverpool on Tuesday at three. . . . I left Liverpool at half-past ten on Thursday morning, and arrived
in London at half-past nine at night, and being at a
loss what to do with myself, went to the nearest hotel,
viz. the Victoria, Euston Square, from whence I immediately wrote to Murchison announcing my arrival.
I did not hear from him till next day (Friday) at five
o'clock, and in the meantime went and saw St. Paul's
and the outsides of some of the streets, for you see
I had always to be running home to look for a letter.
He asked me to breakfast with him on Saturday
morning. This I did. His house is a splendid one.
They are quite people of fashion, but, notwithstanding,
Mrs. M. is a kindly body, and made me quite at
ease at once. I should previously have informed
you that Mr. M. told me in his note that he had
given up the idea of taking me to Russia with him,
but said he was almost certain he had procured me
a much better place, viz. that of Assistant Geologist
to De la Beche, who is at present making the
Ordnance Geological Survey for Government. To
cut the matter short, I may here tell you that on
that day he again wrote to De la B. that the matter
might be finally settled, and on Tuesday last had a
most satisfactory letter from De la B., enclosing
one for me, officially appointing me to the situation
of Assistant Geologist, with pay of 9s. a day. 'Here's
a start.' On Monday first I leave this for Bristol *p*.
Great Western Railway, and on Tuesday I shall be
at Tenby, Pembrokeshire, South Wales, there to join
De la B. Tenby lies, I think, at the mouth of

HENRY T. DE LA BECHE

Milford Haven, a place celebrated by Shakespeare
in *Cymbeline*.[1]
Before leaving, Murchison asked me to dine with
him next day at seven. Mrs. M. also asked me to
breakfast, and to go to church with her afterwards.
The remainder of Saturday I spent getting into my
lodgings, going through the Geological Museum at
Somerset House, calling on Lyell and Graham,[2] and
seeing the Polytechnic. Lyell and Graham both
received me very kindly, indeed Lyell as much so
as Graham. He was very glad to hear of my success,
and told me to be sure and let him know when my
Geology of Arran came out, as he wished to notice
some of my remarks in a new edition of his *Elements
of Geology*. Here's another start. I went to Covent
Garden on Saturday night, and was delighted with
The Critic. On Sunday I went with Mrs. M. in her
carriage to St. Luke's, Chelsea, and having keeked
through the rails and seen the Duke of Wellington,
I went to Westminster at three. At seven I went to
Murchison's to dinner, and there met Mr. Featherstonhaugh, the American plenipotentiary, his lady,
and two gentlemen — a Captain Pringle and Mr.
Munro. Featherstonhaugh is a lively man, but takes
no wine for his stomach's sake.

Monday I spent in the National Gallery and the
British Museum, and in the evening called on Dr.
Stanger, with whom I was acquainted at the meeting.
I found him out by the merest chance. He took me

[1] The writer's literary memory was here better than his geography. Tenby
lies about 18 miles due east from the entrance to Milford Haven.

[2] Thomas Graham, born 1805, died 1869, one of the most distinguished
chemists of our time, was for some years Lecturer on Chemistry in Glasgow, and
in 1837 became professor of the science at University College, London, an
appointment which he held until 1855, when he was made Master of the Mint.
He had known Ramsay and his father in Glasgow, and was one of the first men
of science to welcome him to London.

with him to a Philosophical soirée at Mr. Bowerbank's, and we had a good deal of interesting discussion. On Tuesday, after writing to Nichol and home, I went to Belgrave Square, and there got my official appointment. De la B.'s letter is a very kind one. In his note to Murchison he speaks of my pay rising. I am thoroughly convinced that this is a much better thing than going to Russia. If I behave, and am found worthy, I am sure to rise in the service.

By and by the Survey will go to Scotland. Probably I may get the neighbourhood of Glasgow to do, including my own island. After leaving Murchison I went through Westminster, and saw Dr. Johnson's and Garrick's gravestones side by side, and all the others. 'O rare Ben Jonson!' 'The cloud-capt towers!' I afterwards met Murchison at Somerset House.

Yesterday I spent in the Zoological Gardens, Regent's Park, and also visited the Colosseum. At six I dined with the Geological Club[1] at the Crown and Anchor, Strand. It has a most shabby outside, but is one of those old-fashioned splendid inns inside, which, I suppose, are not to be found out of London. It was here that Fox and the great Whigs of that great day used to meet and enjoy themselves. Lyell and Featherstonhaugh were there, and Captain Pringle; Murchison in the chair. There were about twenty-five gentlemen present. I was introduced to Dr. Buckland and some others. Murchison introduced me also to Mr. Taylor, the croupier and treasurer of the Society, and asked him to take me beside him. I heard him say to Buckland : 'You remember young Ramsay, who made the model of Arran ? I shall intro-

[1] See *postea*, p. 121.

duce him to you.' 'Oh yes,' quoth the Doctor. So I was introduced, and the Doctor gave me two of his digits to shake. There were a lot of big-wigs there whose names I do not know—members of Parliament and others. Mr. Taylor, whom I sat next, knows, or knew, Dr. Thomson of Glasgow, Dr. Ure, Charles Mackintosh, C. Tennant, and others, who were old friends of my father's, and we had a great deal of conversation together. After dinner we went to Somerset House to hear Murchison on Russia. The Marquis of Northampton was there. The discussion broke up about eleven, when we all went upstairs to tea.

I must now close, as I have to go to Belgrave Square and elsewhere, to get my equipment before leaving for Wales.

From De la Beche's letter, containing the formal offer of the appointment, a few sentences may be quoted. It is dated from Cardiff, 22nd March 1841 :

'My friend, Mr. Murchison, having recommended you to me as well qualified to assist on the Ordnance Geological Survey, as I have little doubt, judging from your labours in the Isle of Arran, is the case ; and Mr. Murchison having also stated that you were desirous of joining the service as Assistant Geologist, I have now to offer you the situation of Assistant Geologist on this Survey, with a rate of pay, for the present, of 9s. per day for the six working days of the week (it being the somewhat singular rule that the Sundays are unprovided with pay), payable quarterly, which is at the rate of £140 : 8s. per annum. Independently of this salary, your travelling expenses from station to station would be paid, and all necessary

instruments, drawing materials, etc. etc., are found by
Government.

'Should you feel disposed to join the Survey on
these terms, I would thank you to write to me to that
effect, directing your letter to me, Tenby, South
Wales, to which place I intend to remove my headquarters to-morrow. In that event, it would be desirable that you should report yourself at Tenby on the
1st of April, the commencement of one of our official
quarters. A steamer leaves Bristol for Tenby on
Tuesday, the 29th instant, so that you would only
remain a day or two at Tenby without your pay.'

Though Murchison's strong recommendation may
have had some influence in determining the offer of
this appointment to the young geologist, it must be
remembered that De la Beche had attended the
Glasgow meeting of the British Association, where,
as one of the vice-presidents of Section C, he had
met Ramsay, seen his map and model, and been able
to form an independent judgment as to his capacity
for the work of the Geological Survey.

The pecuniary prospects set forth in the Director-
General's letter could not be regarded as specially
inviting. They were much canvassed in Glasgow,
where the news that Ramsay was not to go to Russia
after all, but had been offered, and had accepted, a
post in the Geological Survey of this country, fell like
a thunderbolt in the family a few days after he had
left home. Mrs. Ramsay's first feeling was one of
bitter disappointment, and it needed all Professor
Nichol's powers of persuasion to convince her that
the situation now offered to her son might really open
the way to his future distinction.

At length, having completed his outfit in London,

he started on the last day of March, and arrived at Tenby at one o'clock in the morning of the 2nd of April by the *Phœnix*, from Bristol, there to begin a career in the Geological Survey which was to last until he had risen to be the head of the service, and one of the foremost geologists of his day.

CHAPTER II

In the Geological Survey of Great Britain, to which Ramsay was now appointed, he spent more than forty years. The work of his life was so intimately bound up with the progress of the Survey that it cannot be intelligently followed unless this relationship is clearly understood. At the outset, therefore, it will be desirable to trace the origin and development of the organisation of which he now became a member, and of which for many years he was the guiding spirit.

The Geological Survey owes its existence to the sagacity and energy of Henry Thomas De la Beche. This distinguished man—the last male representative of a family of Norman barons who came to England with the Conqueror—was born in 1796. From his father, who was in the army, he inherited some landed estate in Jamaica. But the halcyon days of this island had fled, and left him by no means wealthy. It was at first intended that he should follow the profession of his father, and with that end in view he was sent to the Military College of Great Marlow, where he had been preceded some five years earlier by his future friend Murchison. But the close of the great war seeming to shut out any hope of distinction in the career of an active soldier, he turned his thoughts in another direction. From early years he had been fond of natural history pursuits, and especially of geology, for the prosecution of which he had found admirable opportunities along the southern coast of England. When only twenty-one years of age he entered the Geological Society, and two years later was admitted into the Royal Society. He did not confine his attention, however, merely to English geology, but extended his acquaintance with the principles and illustrations of the science by foreign travel. At one time he might be seen sounding and charting the Lake of Geneva, at another he was at work among the rocks on the Riviera, or studying the fossil plants of the Col de Balme. He even carried his science across the Atlantic, and while visiting his paternal domain in Jamaica, lost no opportunity of studying the geology of that island, of which the first account was published by him.

De la Beche had a singularly wide and firm grasp of geological science. A master of stratigraphy, he likewise made himself familiar with minerals and rocks, at a time when the study of petrography can hardly be said to have existed in this country. Having read much and critically in chemistry, he was able to apply the results of chemical research to the problems presented by his geological work. Though not a professed palæontologist, he had such keen sympathy with natural history inquiries, and knew so much of the natural history of his own country, that he recognised from an early period the necessity of a knowledge of organic remains in geological research, and did all in his power to foster the study and applications of palæontology. Moreover, he wrote a number of papers and books. Among his most important and successful works were an excellent *Geological Manual;* a practical treatise, *The Geological Observer,* full of the experience of many years of field-observation; and a striking little volume, *Researches in Theoretical Geology,* which for sagacious insight and breadth of view was far in advance of its time.

To his scientific qualities were added those of the artist and the keen lover of nature, combined with a strength of frame which, in his prime, made him a bold swimmer and an active pedestrian. Over and above all shone his bright cheery nature, his irrepressible merriment, his helpful sympathy, and that inexhaustible enthusiasm which not only supported his own untiring efforts, but, like a contagion, affected and stimulated all who were associated with him. In his later days he was sometimes thought by his officers to be too scheming and to subordinate their interests to the advancement of the Survey and of the large Museum and School of Mines which grew out of it. But we must remember that he had to create the whole establishment, to gain the goodwill of successive governments and ministers, not always predisposed to spend money in the cause of science, and to keep the organisation effective with as little outlay as possible.

After various more or less desultory geological studies at home and abroad, De la Beche at last settled down seriously to the detailed investigation of the geological structure of the south-west of England. He began to map that region on the Ordnance maps which had then been published on the scale of one inch to a mile. He soon saw of how much practical utility carefully-prepared geological maps would be in aiding the development of the mineral resources of the country, and that the work which he was himself voluntarily undertaking at his own charges would be more efficiently performed in connection with the general Trigonometrical Survey of the country. Accordingly, having laid his views before the authorities, he was in 1832 appointed by the Board of Ordnance 'to affix geological colours to the maps of Devonshire and portions of Somerset, Dorset, and Cornwall.'[1] His work thus obtained official recognition.

By the beginning of 1834 De la Beche, acting under the direction of the Board of Ordnance, had produced a geological map of the county of Devon which, as remarked at the time by Greenough, 'for extent and minuteness of information and beauty of execution has a very high claim to regard.'[2] He worked with such rapidity that by the end of that year, of the eight sheets of the Ordnance map on which he had been engaged, four had been published, three were complete, and the eighth nearly complete, while the explanatory memoir and sections were far advanced.[3]

Next year (1835) an important step was taken in the official recognition and assistance of De la Beche's labours. Owing, no doubt, to his own representations on the subject, the Ordnance authorities were led to consider the question of the geological work which he had been carrying on under their sanction, and to take the advice of distinguished experts in regard to it. Their action and its results cannot be better told than in the following quotation from the address of Lyell as President of the Geological Society in February 1836.

Early in the spring of last year an application was made by the Master-General and Board of Ordnance to Dr. Buckland and

---

[1] *Proc. Geol. Soc.* i. p. 447.   [2] *Op. cit.* ii. p. 51.
[3] *Op. cit.* p. 154.

Mr. Sedgwick, as Professors of Geology in the Universities of Oxford and Cambridge, and to myself, as President of this Society, to offer our opinion as to the expediency of combining a geological examination of the English counties with the geographical survey now in progress. In compliance with this requisition we drew up a joint report, in which we endeavoured to state fully our opinion as to the great advantages which must accrue from such an undertaking, not only as calculated to promote geological science, which would alone be a sufficient object, but also as a work of great practical utility, bearing on agriculture, mining, road-making, the formation of canals and railroads, and other branches of national industry. The enlightened views of the Board of Ordnance were warmly seconded by the present Chancellor of the Exchequer (Mr. T. Spring Rice), and a grant was obtained from the Treasury to defray the additional expenses which will be incurred in colouring geologically the Ordnance county maps. This arrangement may be justly regarded as an economical one, as those surveyors who have cultivated geology can with small increase of labour, when exploring the minute topography of the ground, trace out the boundaries of the principal mineral groups. This end, however, could only be fully accomplished by securing the co-operation of an experienced and able geologist, who might organise and direct the operations; and I congratulate the Society that our Foreign Secretary, Mr. De la Beche, has been chosen to discharge an office for which he is so eminently qualified.[1]

The amount granted by the Treasury was only £300 a year, so that most of the expense of the surveying still fell upon De la Beche himself. He obtained, indeed, occasional assistance from two officers of the Ordnance Survey[2] who possessed some geological knowledge, and who more especially helped him in the mining districts.

At last by the year 1839 all the maps of the southwest of England had appeared ; likewise an admirable octavo volume, giving a description of the geology of this interesting and important region.[3] How these

---

[1] *Proc. Geol. Soc.* vol. ii. p. 358.

[2] H. M'Lauchlan and H. Still, who were both Fellows of the Geological Society.

[3] *Report on the Geology of Cornwall, Devon, and West Somerset,* by Henry T. De la Beche, F.R.S. and Director of the Ordnance Geological Survey, 1839. The Survey maps are sheets 20-33.

publications were regarded at the time by English geologists may be gathered from the encomium pronounced on them by Buckland as President of the Geological Society in the spring of 1840.

The first map which I shall mention affords another example of the recognition by Government of the importance of our subject by their having attached a geological department to the Ordnance Survey of England and Wales. The first-fruits of this appointment are the splendid maps of Devon and Cornwall and a part of Somerset, coloured after the surveys of Mr. De la Beche; and it may truly be said of them that they are more beautiful in their execution, more accurate in their details, and more instructive in the economical and scientific information they give respecting mines than any maps yet published by any Government in the world; affording documents to which we can at length with pride appeal, in reply to the reproach that has so long, with too much truth, been cast upon us, that England alone, of all the civilised nations, has abandoned to gratuitous individual exertions, and the liberality of amateurs in science, the great work of exploring and delineating the mineral structure of the country, and ascertaining the nature and extent of the subterranean produce which lies at the foundation of the industry of its manufacturing population, and to which the nation owes no small portion of its wealth.[1]

The rapidity with which these maps were prepared by so small a staff would have been impossible had the ground been surveyed in the same minute detail as is now practised. In fact, admirable as they were in many ways, and far as they were in advance of anything of the kind previously attempted, they can be regarded as little more than sketch-maps, giving a first general outline of the geological structure of the ground. They ought not to be judged by the higher standard of intricate detail subsequently developed in the work of the Survey. In later years, had Ramsay been free to act as he pleased in the matter, he would have had all these early maps resurveyed.

Having so successfully launched his scheme for a

[1] *Proc. Geol. Soc.* iii. p. 221.

geological survey of the kingdom, De la Beche proceeded to point out to the Chancellor of the Exchequer
that for the adequate development of the great mineral
industries of the country it was not enough to make
accurate maps of geological structure, but that it was
further needful to collect and exhibit specimens of
rocks and minerals which were used, or might seem
capable of application, in the industrial arts. He had
already, during his work in Devon and Cornwall,
made an extensive collection of specimens from the
great mining region of the south-west. Another large
series of samples of British building-stones was ac-
cumulated by the Commission appointed to inquire
into the most suitable materials for constructing the
new Palace of Westminster, after the burning of the
old Houses of Parliament in 1834.[1]

There was thus a large amount of material ready
for display, and through the labours of the Ordnance
Geological Survey, as well as from donations, it was
continually increasing in extent and in value. De
la Beche's representations were so obviously well
founded, that they soon obtained official approbation. Apartments were allotted for the accommodation of the Survey collections, and in February 1837
the Office of Woods and Forests formally took the
scheme under its charge, and asked De la Beche to
carry out his proposals. His design was to establish
a Museum of Economic Geology, wherein the

---

[1] This Commission consisted of Mr.—afterwards Sir Charles—Barry, William
Smith, the father of English Geology, De la Beche, and Mr. C. H. Smith, a
practical sculptor. De la Beche probably took the main part of the labour of
collecting the specimens and preparing the Report. The work was done before
the days of railroads, and the Commissioners drove about the country in an old
carriage and pair, visiting quarry after quarry, procuring rough samples of the
different stones, which were sent up to Mr. C. H. Smith's yard to be dressed into
six-inch cubes. These blocks are now in the Museum of Practical Geology,
Jermyn Street.

various practical applications of the science might be thoroughly illustrated by specimens, models, maps, sections, and as much information as possible, not only for the general public, but especially for the guidance of all persons practically interested in mineral substances and their applications.

The premises assigned to him for the housing of his collections were in a plain building of moderate size, with no front to the street, and situated in the retired space known as Craig's Court, Charing Cross. The Museum of Economic Geology, thus started, was in fair working order by 1839, though not ready to be opened to the public for two years later. It was under the control of the Office of Woods, but the Geological Survey remained as a branch of the Ordnance Survey, and De la Beche directed the Museum gratuitously. So vigorously did he set to work that, besides the specimens of rocks, minerals, and fossils, he soon gathered together, arranged, and displayed models of mines, samples and models of mining machinery and apparatus, with illustrations of metallurgical processes and of the various industries which arise from the manipulation of mineral substances.

He further secured sanction to fit up a laboratory, and to appoint as Curator of the Museum one of the best analytical chemists of his day, Richard Phillips, who had taken part in the foundation of the Geological Society. At this laboratory it was arranged that the public might obtain analyses of rocks, minerals, and soils.

There was yet another important department which De la Beche now organised. The British Association had in 1838 memorialised Government to collect and preserve documents recording the mining operations of the United Kingdom, on the ground that, for want of the proper preservation of such records, great loss of life and destruction of property had taken place. This petition having been favourably received, De la Beche was authorised to form a Mining Record Office as part of the Craig's Court establishment. Plans and sections of mines were obtained from various mining districts, steps were taken to procure statistics of mineral industry, and in 1840 the Mining Record Office thus started was committed to the charge of T. B. Jordan, a man of remarkable ingenuity, who had been Secretary of the Royal Polytechnic Society of Cornwall. It was further arranged that lectures should be given on the subjects illustrated by the Museum.[1]

When he had completed, with so little aid, the survey of the south-western counties, and had roused the Government of the day to some appreciation of at least the industrial value of his work, De la Beche resolved to transfer his field-operations to South Wales, where an important coal-field awaited examination. Still under the Board of Ordnance, he obtained increased parliamentary grants, and was allowed the services of a few assistants—young men with no geological experience, whom he had to train in all the details of geological mapping.

The field-work had been a year or two in progress in South Wales when Andrew C. Ramsay joined the staff. Referring to this period of his life at a much later time, he remarked : ' In the year 1841 I had the good fortune to be appointed one of the few assistant geologists. The Survey had then progressed west-

---

[1] For the early history of the Museum of Economic Geology and Mining Record Office see the *Account* of them by T. Sopwith (see p. 78), published by Murray in 1840. See also Buckland, *Proc. Geol. Soc.* iii. (1840), pp. 211, 221. The Mining Record Office was transferred to the Home Office in 1883.

wards into Pembrokeshire, and was at work at Tenby, and St. David's, and the neighbourhood. There were then four assistants besides myself.'[1]

Over and above the ordinary assistants, however, the Survey was aided in the palæontological department by Professor John Phillips—a name affectionately remembered by those who knew him, and honoured by all to whom the history of British geology is familiar.[2] Phillips had previously been employed to examine, figure, and describe the organic remains in the older rocks met with in the course of the survey of Cornwall, Devon, and West Somerset, and an important monograph giving the results of his labours appeared in 1841 as a sequel to the *Report* of De la Beche.[3] Before the publication of this work, however, the field-operations of the Survey had extended into South Wales, and Phillips in 1840 received an appointment to extend his task into East Somerset, Gloucester, Monmouth, and South Wales. He was in Pembrokeshire when Ramsay joined the staff, and they had some excursions together.

The duties of the geologists in the Geological Survey were to trace on the one-inch maps of the Ordnance Survey the boundaries, structure, and relations of the various geological formations, to collect as much information as possible regarding the nature of the rocks and minerals, to mark where any substances of economic value might be found, to follow

---

[1] 'On the Origin and Progress of the Geological Survey of the British Isles' in *Conferences held in connection with the Special Loan Collection of Scientific Apparatus*, South Kensington Museum, 1876, p. 364. The four assistants referred to above were W. T. Aveline, who joined the service the year before Ramsay, retired from it the year after him, and now lives in Somerset ; Trevor E. James, D. H. Williams, and J. Rees.

[2] See *ante, footnote*, p. 18.

[3] *Figures and Descriptions of the Palæozoic Rocks of Cornwall, Devon, and West Somerset.* By John Phillips, 1841.

out the crops of lodes and mineral seams, as well as of
the more important dislocations in the crust of the
earth, to note where fossils occurred, and to take such
specimens of minerals, rocks, and fossils as might be
required for the preparation of the maps for the
engraver, and the compilation of material for the subsequent explanatory Memoirs. There were likewise
levellings to be executed for the purpose of constructing
horizontal sections, which were drawn on the scale of
six inches to a mile. These sections formed as novel a
feature as the detailed maps in the progress of geological surveying. They had been constructed by Logan
in Wales, in order to represent accurately the structure
of the great South Welsh coal-field. The same scale
was adopted by De la Beche, who, with his artistic eye
and deft hand, introduced into his horizontal sections
a system of representation of geological structure such
as had never before been attempted. The sections
were on a true scale, vertical as well as horizontal.
By carefully chaining and levelling, the topography of
the ground was represented correctly, and for the first
time the relations between surface-features and under-
ground structure were clearly brought out.

In carrying out the various field-operations of the
Survey De la Beche took an active personal interest.
He spent the greater part of the year with his officers,
and kept himself in touch with the details of their
work, besides continuing for some years to carry on
independent mapping of his own. As the staff increased in number, and Ireland came under his
jurisdiction, he was necessarily prevented from doing
much himself in actual mapping, and he gradually left
more and more to the judgment of his subordinates.

Being under a military organisation, the surveyors

WILLIAM E. LOGAN

wore a dark blue uniform. A tight-fitting, well-buttoned frock-coat, however, was not a very comfortable garment for the rough scrambling and climbing work of the survey life. The geologists were therefore by no means sorry when, on their transference in 1845 from the Ordnance Department, they were at liberty to choose their own civilian apparel. But as a souvenir of their military connection they retained the gilt buttons embossed with the crown and crossed hammers, which for many years afterwards were worn on festive occasions. Even those members of the service who joined in later years used to provide themselves with a set of the 'Survey buttons,' and wore them on their waistcoats at the annual dinner.

The life of a member of the Geological Survey is, in many respects, an enviable one. He starts soon after breakfast, lightly accoutred, and spends the day, map in hand, over the ground assigned to him for survey. Every exposure of rock is noted by him on his map or in his note-book, with all the needful details. Each stream is followed step by step up to its source ; each hill-side and ravine is traversed from end to end ; each quarry, sometimes each ditch, and even the very furrows and turned-up soil of a ploughed field are scrutinised in turn. He is thus led into every nook and corner of the ground, until he acquires a more intimate knowledge of it than many of the natives who have been living there all their lives. Out early and late, and in all kinds of weather, he witnesses changing atmospheric effects such as few others have opportunities of enjoying. He is brought into every variety of scenery, and is compelled by his very duties to study these varieties, and make use of them in his daily work.   If he has a love of nature, and this, to be a good geologist, he must possess, he is afforded ample scope for its gratification. Flowers, insects, birds, and living things of every kind meet his eye at each turn of the way.   If he has any antiquarian instincts, his rambles enable him to visit every antiquity for miles around him.   If, lastly, he is of a social temperament, and cares to mix with his fellowmen, there is often pleasant society in the neighbourhood, where a stranger of good address is generally welcomed.   Sometimes he must content himself with the kindly gossip of the little farm or way-side-inn ; at other times he finds himself discussing rural politics with the village doctor, or undergoing a process of examination in the tendencies of modern science at the parsonage, or joining in a pleasant dinner-party at the squire's.

That such a life has its trials, however, may readily be believed.   The mere physical endurance which it often requires is enough to tax the strength of a strong man.   Not unfrequently, indeed, it involves personal danger as well as discomfort.   Few members of the staff but can give instances of narrow escapes from fatal accident.   Now it is a mass of cliff or crag which, without warning, falls with a crash close to where the surveyor is standing, or a single loosened block from the rocks above shoots past his head with a whizz like a cannon-ball.   At another time it is a treacherous bog which, firm apparently on the surface, suddenly gives way under his feet, and out of the mire of which he with difficulty drags himself.   Streams which in the morning could be jumped across may by nightfall, after heavy rain, be so swollen as to be unfordable without peril.   Snow - storms sometimes surprise the geologist among the hills, and as the snow rapidly gathers, roads, walls, and fences may be entirely buried before he can struggle through the blinding drift back to his quarters. Among the mountains he is apt to be overtaken by mists so dense that much skill may be needed to steer a right course through them. And in thunderstorms he is sometimes startled by the lightning flash which strikes a tree or a house, or kills a cow, quite close to him.

But apart from occasional personal risk, the constant exposure to the vicissitudes of a changeable climate, the necessity of sometimes enduring serious discomfort and privation in districts where quarters are hardly to be had, where the food is of the sorriest kind, and yet where the geological work may be most difficult and prolonged; the isolation and loneliness at stations where no congenial society of any kind is to be found, the necessity of frequently moving camp to begin all the domestic experiences and discomforts over again, and the poor pay for which all this drudgery has to be undergone—these and other hardships which may be easily imagined test the scientific enthusiasm of a geologist. By a young man who is fired with an ardent love of his science they are lightly regarded and soon forgotten. It is only as he grows older, and his enthusiasm somewhat wanes, that he begins to find them a serious impediment to the settled home which he then, not unnaturally, longs to establish.

It may easily be imagined that when a member of the Survey plants himself in a country village his occupation becomes at once a source of the utmost curiosity to his neighbours. He carries his accoutrements about his person in such a manner that they do

not attract notice. Compass, clinometer, map, note-
book, lens, pencils, and so forth, are easily stowed
away in his pockets; the hammer can be disposed in a
belt under the tails of his coat, so that he presents no
outward marks of his profession. His movements are
consequently mysterious in the extreme to the villagers
and farm-people, and the most amusing mistakes are
made in endeavouring to guess his calling in life. He
finds himself set down now for a postman, now for a
doctor, for a farmer, a cattle-dealer, a travelling show-
man, a country gentleman, a gamekeeper, a poacher,
an itinerant lecturer, a gauger, a clergyman, a play-
actor, and often as a generally suspicious character.
A former distinguished member of the staff, who now
holds a University professorship, has received and duly
posted many a letter entrusted to him in the belief that
he was the authorised bearer of Her Majesty's mails.
Another well-known colleague, who is now also a
University Professor, tells how on one occasion he
was taken for a policeman in plain clothes, and could
not for some time make out why a poor woman poured
into his ears a long story about her son, who had been
taken up for doing something that he had not done,
and did quite unintentionally, and was quite justified in
doing. Gamekeepers are sorely puzzled sometimes
what to make of the Geological Survey trespasser;
they are afraid to challenge him lest he prove to be a
friend of their master, and afraid to let him go his way
for fear he be on poaching thoughts intent. One
member of the staff who had taken up his quarters in
a village was watched for some days by the police on
suspicion of having been concerned in a recent burglary.
Another was stalked as a suspect who had been setting
fire to farm-buildings. A third was watched hammer-

ing by himself in the bed of a stream, and as he gave vent to some strong expression when the obstinate boulder refused to part with a splinter, the onlooker on the other side of the hedge fled in terror to the neighbouring village and reported that this strange man who had come among them was stark mad, and should not be left to go by himself. Sometimes the laugh goes distinctly against the geologist, as in the case of one of the staff who, poking about to see the rocks exposed on the outskirts of a village in Cumberland, was greeted by an old woman as the 'sanitary 'spector.' He modestly disclaimed the honour, but noticing that the place was very filthy, ventured to hint that such an official would find something to do there. And he thereupon began to enlarge on the evils of accumulating filth, resulting, among other things, in an unhealthy and stunted population. His auditor heard him out, and then, calmly surveying him from head to foot, remarked, 'Well, young man, all I have to tell ye is that the men o' this place are a deal bigger and stronger and handsomer nor you.' She bore no malice, for she offered him a cup of tea, but he was too cowed to face her any longer.

When Andrew Ramsay entered upon this roving Survey life in the spring of 1841, he was twenty-seven years of age, active and athletic in body, with boundless enthusiasm for geology, and an ardent desire to devote himself to practical geological work. Long afterwards, looking back on this period of his life, he used to tell how at first the change from a Glasgow counting-house to daily occupation among the hills and along the shores of South Wales seemed like a dream. He could hardly realise for some time that the pursuit, formerly followed only during brief

E

but coveted intervals of holiday, was now to be the constant business of his life. Day after day, as he went out with map and hammer, it seemed to him still holiday work. He brought the same insight and ardour to the study of the Welsh region as he had shown in that of Arran. And before long the Director recognised that in his new recruit he had obtained by far the ablest member of his staff.

For upwards of four years the Survey continued to advance across South Wales. During this period Ramsay gradually worked his way northwards from the southern coast-line of Pembrokeshire across the counties of Caermarthen, Brecknock, and Cardigan, into those of Montgomery and Radnor. Professor John Phillips was with him for a short time along the Pembrokeshire coast before establishing himself among the Malvern Hills, which he mapped in detail. H. W. Bristow was at work in Gloucestershire, where Ramsay joined him (p. 235), and afterwards, under Phillips, took a share in mapping the Oolites of the Cotteswold Hills and of the Cheltenham, Wotton-under-Edge, and Bath district.[1] But in these early years his time was almost wholly devoted to the older rocks in Wales.

The geological structure of the Welsh region in which he was called upon to labour proved to be excessively complicated. It had been only cursorily examined by previous observers. De la Beche and Phillips were content to map its southern outskirts in a somewhat sketchy manner. Its real difficulties remained to be discovered and grappled with. After his first year's experience Ramsay drew up a draft report of his operations. Unfortunately this report was never printed, nor do its conclusions appear to have been

[1] See Prefatory Notice to ' Geology of East Somerset,' *Mem. Geol. Surv.* 1876.

published even in abstract.   It is entitled 'Report on
the work entrusted to A. C. Ramsay in North Pembrokeshire, and part of Cardiganshire and Caermarthenshire.'   The MS., which is in his handwriting,
remains in the archives of the Geological Survey.   In
this document he gave special prominence to the
igneous rocks, which he separated into intrusive and
contemporaneous, showing that the latter cover by far
the greater area.   Among the rocks of St. David's
he clearly recognised the presence of volcanic ash, and
saw in these rocks the records of prolonged volcanic
activity.   Other geologists, notably Sedgwick, Murchison, and De la Beche, had described the proofs of
contemporaneous volcanic eruptions among stratified
formations of old geological date.   But Ramsay was
the first to trace out in detail the structure of a
volcanic series of such high antiquity, and to separate
from each other the outflowing lavas, the ejected
ashes, and the deep-seated intrusive sills.   When we
remember, too, that this was practically his first piece
of detailed mapping, we cannot fail to acknowledge
the earnest which was thus given of the future
geological accomplishment of the surveyor.

It fell to Murchison's lot as President of the
Geological Society in 1843 to give some account of
the recent proceedings of the Geological Survey in his
address at the Anniversary of the Society in February.
He referred to the increasing evidence brought forward by the officers of the Survey that the interior of
South Wales, which had been vaguely referred by him
to Sedgwick's 'Cambrian' system, consisted largely
of Lower Silurian rocks.   He spoke of the Survey's
'results, obtained among strata so obscured by change,
as among the very highest triumphs of geological

field-work.' ' I therefore wish,' he added, 'to be foremost in recognising the deserts of the labourers who
have obtained them, among whom the Director particularly cites Mr. Ramsay, already so favourably
known to us by his geological map and model of the
Isle of Arran.'[1]

The St. David's map was published in 1845, and
after its appearance a sheet of horizontal sections was
prepared and issued, showing what was believed to be
the general structure of the ground. Unfortunately,
twelve years afterwards, in second editions of these
publications, while great improvements were made in
the general stratigraphy, the views originally formed
by Ramsay as to the nature of the St. David's rocks
were so modified, though confessedly with his own consent and co-operation, that the essentially accurate interpretation at first adopted disappeared. In later years
the truly volcanic nature of much of the fragmental
rocks in that district, which in the second edition of
the map became ' altered Cambrian,' was re-discovered,
and the merit of the first observations was for a time
obscured.[2]

There was yet another feature in which Ramsay
improved the mapping of the Survey. He traced out,
where practicable, lithological subdivisions among the
older Palæozoic rocks, which had not previously been
subdivided, and was thus able to detect their sequence
and the general structure of the ground over which
they extended. In particular, even in his first year's
work, he drew a line between the black and purple
slates which, though not put on the published map at

[1] *Proc. Geol. Soc.* iv. (1843), p. 76.
[2] The details of this question will be found narrated in the *Quarterly Journal
of the Geological Society*, vol. xxxix. (1883), p. 263.

the time, was afterwards adopted as the boundary between the Cambrian and Silurian systems.   It was in those days the belief of the great body of geologists that the older rocks of South Wales belonged to Sedgwick's Cambrian formations, and as such they were coloured on the large map accompanying Murchison's *Silurian System*, published in 1839.   A few fossils had indeed been found in them which were of Lower Silurian species, but the evidence supplied by these fossils does not seem to have been considered strong enough to change the general current of opinion. When in 1841 the Survey began to map the region about Haverfordwest, neither De la Beche nor his officers could find any base to the series which, by common consent, was acknowledged to be Lower Silurian.   And when in that and the following year Ramsay and others obtained Lower Silurian fossils at various points across the whole breadth of South Wales, they could come to no other conclusion than that this wide region consisted of Lower Silurian rocks repeated in endless undulations.

Much more detailed work would now be possible in South Wales than is shown upon the maps of the Geological Survey.   But those who may in future carry out this re-survey will doubtless be the first to admit the value of the work of the pioneers who produced the first geological map of that difficult tract of country.   Ramsay himself was well aware of the imperfection of the early work in South Wales, as will be apparent in later pages of this volume.

Of the actual daily life of these first years of his Survey experience in Wales little record seems to have been preserved.   He used to tell in later life how, when stationed at St. David's during

the summer of 1841, he sang in the Cathedral choir,
for he had an ardent love of music, could with facility
read music at sight, and possessed a good voice.
One of his early experiences he used sometimes to
recount: how, having been benighted among the hills,
he found his way in the dark to a stream-course, and
in descending it came to a cottage where he was
known. The shepherd brought him in out of the
darkness, and his wife, seeing the famished look of the
wanderer, set a large dish of food before him. Eating
with all the 'passion of a twelve hours' fast,' Ramsay
soon emptied the dish, and then to his dismay discovered that he had eaten up the supper of the family.
From his pocket diary for 1842, which has survived,
we get a few further glimpses into his proceedings.
Instead of going up to London he remained at his
field-quarters all winter, and went out among the hills
when the weather permitted. He continued his active
pedestrianism, sometimes covering 30 miles in a day.
When the distances from his station got too far to be
easily reached on foot, he would ride out to his
ground, put up his horse at a farm, spend all day in
mapping, and ride back to his quarters in the evening.

On wet days and in the evenings he had always
plenty of occupation indoors. He was a regular correspondent with his family in the north, and with many
of his old Glasgow companions, now scattered over the
world. In the brief jottings of his memorandum books
he always inserted the names of those to whom he had
written. Hardly any of these early letters have been
recovered. Besides writing to his mother and sister,
who were now all that remained in the old home, from
the very beginning of his Survey life he remitted
money out of his income to them, and he continued

this pious duty as long as his mother lived and his sister remained unmarried.

Of an eminently social temperament, he made acquaintances easily wherever he went, and these chance acquaintanceships sometimes ripened into life-long friendships. In one family circle we find him reading aloud Shakespeare, or Scottish ballads, or a good novel ; in another he takes part, heart and soul, in singing glees and madrigals ; in a third he joins in dancing and all kinds of merriment. After being some little time at a station he knew everybody worth know-ing all round him, and sometimes had difficulty in satisfying the demands for his company. The appear-ance of a pleasant, conversational, and merry-hearted stranger was sometimes an extraordinary boon to a country district in the days before railways. His doings and sayings, his goings-out and comings-in, were a source of the deepest interest to gossips who longed for some new event in their little world. If he dined at a house noted for its conviviality, there would be solemn head-shakings and expressions of regret that one so young should have been led into such courses. If he spent an evening now and then in a family where there were two or three daughters, it was supposed that he could have but one object there. Curiosity was at once aroused to discover which of the ladies he had chosen, and curiosity soon gave way to certainty as the report of an engagement was rapidly circulated through the parish.

In South Wales Ramsay had early experience of these manifestations of public interest in his affairs. He was naturally fond of female society. His conver-sational powers, his literary taste, and his lively humour found there a congenial stimulus. And while he thoroughly enjoyed it, his presence brought a brightness which gave general pleasure in return. This mutual reaction continued to mark his social intercourse up to the end. In his younger days in Wales, when the tittle-tattle of a countryside was beginning to teach him greater circumspection, he passes judgment rather severely on himself. '13*th* *February* 1842.—Am, on the whole, rather an ass to be so serio-comic, sentimental, and universally *captivato flirtaceous.*'

Although most of the time working alone, he had occasionally visits from one or other of his colleagues, or went to see them at their stations. De la Beche, too, used to join him and spend a few days with him on his ground. The Director-General was knighted in April 1842, and early next month paid a visit to his officers in South Wales. The following entries occur in the diary of 1842 : '3*rd May*.—Sir Henry De la Beche arrived at Caermarthen. I called at night, when we walked down to the stables to see his horse. As jolly as ever.' '4*th*.—Out all day with Sir Henry.' '13*th*.—Sir Henry left me to-day.' '9*th* June.— To-day we [A. C. R. and T. E. James] found lots of fossils far to the north of Llandeilo; wrote to Sir H. to come and see them.' '10*th*.—Had a glorious find of fossils, and played at cricket in the evening.' '11*th*. —Sir Henry came down and saw our beautiful section by Cwm y Wern. We all bathed in a pool. Fossilised a bit, and then home.' '14*th*.—Rode to Llangadock to see Rees; a delightful night; Sir H. very kind about futurity.'

It was one of De la Beche's characteristic traits that, having wide aims, and clear views as to how he should endeavour to carry these aims to their fulfilment, he possessed a power of discerning the qualities of the fellow-workers who would best serve his purpose, and of attracting and attaching them to his corps. He was always moving about with his eyes open, on the outlook for the best men to co-operate with him in his great scheme for the national endowment of geological inquiry. After many months of delay and suspense he succeeded, in the autumn of 1844, in inducing the Government of the day to authorise him to appoint a palæontologist to the staff of the Survey, whose special duty it should be to determine the fossils found by the surveyors and collectors, and to confer with these officers in the field as to the classification and boundary lines of the fossiliferous formations. He selected for this important post perhaps the most brilliant naturalist of his day—Edward Forbes, who, after returning from his researches in the Ægean Sea, had been appointed Curator of the Geological Society. Ramsay and Forbes had formed a friendship at the Glasgow meeting of the British Association, and their intimacy grew every year closer. The subjoined letter is of interest here.[1]

22nd November 1844.

DEAR RAMSAY—What on earth put it into your head that I could possibly be offended at anything you have lately written and done? My dear fellow, I feel most grateful to you, and it is only the press of business engagements upon my change of office which has prevented my writing. I can assure you I feel 10° happier than I did last winter and spring, having now a fair promise of doing something satisfactory for science, and getting rid of Geological Society patchwork. In a fortnight I shall be altogether clear of the G. S. (as an officer, that's to say), as I suppose they will make their election next meeting. Ansted [see p. 78] will probably be the man, and a better for their purpose they could not have. He'll keep up the dignity of the office, and work like a brick. . . . I look forward to

---

[1] The triangle at the end of Forbes's signature was characteristic of his early days. See his *Life*, p. 195.

great things at C. C. [Craig's Court], and have the fullest faith in Sir Henry.

As to news here, there isn't much stirring yet.   It may be divided into Scientific, Literary, and Philosophical—

1. *Scientific*.—Mantell is giving a course of geological lectures at the London Institution.   A curious book called the *Vestiges of Creation*, containing many speculations on Geology and Natural History, said to be written by Sir Richard Vivian, is making great stir in town.   A first-rate synopsis and analysis of the Trilobites has just come to hand from Germany with exquisite plates.   It is written by the great entomologist Burmeister.

2. *Literary*.—*Punch* is published as usual every week.   Two new plays have just come out, and the theatres are worth going to.

3. *Philosophical*.—Lankester, Day, Francis, Henry, etc. etc., and myself have succeeded in establishing a monthly meeting and feed of Red Lions at the 'Cheshire Cheese' in Fleet Street [see p. 62].   We look forward to your roaring in our company.   Clara ——, the pretty dancer, ran away with somebody the beginning of last week, and came back at the end of it.   Amen !   Ever, dear Ramsay,

EDWARD FORBES Δ.

Ramsay and Forbes in the early part of this year (1844) had been sounded as to the feasibility of bringing out, with the sanction and co-operation of W. D. Conybeare,[1] a new edition of the classic *Outlines of the Geology of England and Wales*, by Conybeare and W. Phillips, of which the original edition appeared in 1822.   Conybeare himself had made some progress with the task, but he seems to have found the labour beyond the strength of his advancing years, and through the intermediary of De la Beche, entered into negotiations with the two younger authors.   These proposals took at last definite shape in a formal legal agreement, dated 8th April 1844, between Conybeare, Forbes, Ramsay, and Messrs. Longmans and Co., publishers.

[1] William Daniel Conybeare, born 1787, died 1857, author of some important geological memoirs, but best known for the *Geology of England and Wales*, referred to above.   This work was properly the second and much enlarged edition of a volume by W. Phillips, which was published in 1818.   It did not include an account of the older Palæozoic rocks.

The work was to be in three parts or volumes, of which it was stipulated that the first should be delivered complete by the 1st January 1845, the second by the 1st October of the same year, and the third by the 1st October 1846. In fulfilment of this undertaking Ramsay made numerous notes, and wrote out many pages of manuscript, but the press of official and other engagements, which became, both with him and with his colleague, increasingly severe, prevented the task from ever being completed. Looking back upon this enterprise, we can hardly doubt that Ramsay felt his practical acquaintance with English geology to be as yet too limited to enable him to perform the task as he would have wished. It was only three years since he had left Scotland, and those years, actively spent in field-work as they were, had been passed almost wholly in South Wales. He gained eventually an unrivalled familiarity with English geology, but many years had still to elapse before that qualification was acquired.

Much is said and written in dispraise of the climate of the British Isles, but the field-geologist can find few regions on the face of the globe where he may ply his vocation more continuously from season to season than there. The winters over much of the United Kingdom are seldom so severe as seriously to interrupt out-of-door work for more than a week or two at a time. The summers are not too warm to prevent active exercise in the open air from early morning until dusk, while the length of a summer's day in these northern latitudes gives time for as much continuous walking and climbing as the strongest frame can endure. Even the rain, which is the geologist's chief meteorological enemy, falls in such wise that the number of thoroughly wet days, when nothing can be properly accomplished out of doors, is much less than most people would be apt to believe. Hence, even as far north as central Scotland, it is quite possible to carry on geological surveying throughout the whole year. And often the clear bracing air of December allows nearly as much work to be done in a day as can be accomplished in the warmer but more exhausting weather of June. Accordingly, it was no hardship to Ramsay that, for the first year or two of his Survey life, he spent the winters in South Wales. Thereafter he generally came up to winter quarters in London, the building in Craig's Court serving as the head office of the Survey.

As occupation for the members of the Survey during the winter months there is generally a considerable accumulation of indoor work which cannot be satisfactorily completed in country quarters. The lines traced on the field-maps have to be drawn on fresh copies, or what are called 'dry-proofs' of the sheets, and all the details must be inserted which are intended to be published, preparatory to the engraving of the work. The horizontal sections levelled in the country have to be plotted to scale, and their geological details to be inserted. There are likewise reports and descriptions which require to be extended from the field note-books. There is thus usually ample occupation to keep the surveyors busy from the time when they drop field-work towards Christmas till they resume it in spring.

In the early days of the Survey's history most of the staff were young and unmarried. They took lodgings in London, and generally dined two or three or more together in some restaurant. Once a fortnight came the meeting of the Geological Society, where they usually made their appearance, and, seated on the back benches, looked down upon the veterans on the front rows, and listened to the papers and discussions, often lively enough in those early days of geology. They had the entry also into various social gatherings, with an occasional night at the theatre, so that the time they could secure for quiet reading was by no means great. How Ramsay passed his time in the first seasons of his London life may be gathered from a few extracts from his diary of the early months of 1845 :—

'*3rd January.*—Reached the Paddington Station [from Wales] at five in the morning. Got down to the "Golden Cross," slept on benches, and breakfasted at eight. At ten met Sir H. at the Muzzy [Museum], and had a most jolly reception. Minute from the Treasury authorising the junction of the Survey and the Muzzy. It is proposed also that I should be Sir H.'s first lieutenant with £300!!! Thus one dream is in a fair way of being realised. Playfair and I dined and then danced at Smyth's till four.'

The change in the official relations of the Survey thus briefly alluded to was a momentous one in the history of the service. It was now arranged that the Survey, hitherto conducted under the Board of Ordnance, should be transferred to the Office of Works, and that the Museum and Survey should thus be united as part of one organisation under the control of a single public department. It was further provided that the staff of surveyors should be increased ; that the Geological Survey of Ireland, which had likewise been in charge of the Board of Ordnance, should henceforth be placed under

the supervision of the Director of the Geological
Survey of Great Britain, who should assume the title
of Director-General, and that under this chief there
should be two Directors, one for England and Scotland, with the title of Local Director for Great Britain,
and one for Ireland. It was the former of these two
posts which was now to be conferred on Ramsay.

'*8th January.*—Playfair and I dined together.
Geological Society night. Paper on Fossil Crania
from South-Eastern Africa, by Owen. He, Forbes,
and I supped with Playfair after. Owen's genius
throws light on everything.

'*16th.*—Dined for the first time with the Metro-
politan Red Lions in my capacity as a corresponding
member. Smyth and Falconer elected members.'
This fraternity took its rise at the Birmingham meeting of the British Association in 1839. Edward
Forbes and a few congenial spirits, finding the dull
conventionality of the ' ordinaries ' insupportable,
started simple dinners of their own, where beef and
beer were the chief viands, and where the mirth and
jollity were so great that admission to these gatherings
soon came to be eagerly sought after. The place of
meeting was a modest inn known as the ' Red Lion,'
and the company styled themselves therefrom ' Red
Lions.' They agreed to meet at every meeting of the
British Association—a custom which they and their
successors have kept up till the present time. Those
of them who lived in London, with Forbes at their
head, feeling that a year's interval made too wide a
gap between their festive gatherings, formed themselves into a London company of the original brotherhood, and it was to the monthly dinner of this company that Ramsay was now introduced.

'*19th January.*—Home to read [Hugh Miller's] *Old Red Sandstone.*

'*19th February.*—Drew [sections]. Phillips (John) came, and we had a big talk. He is still to join us for six months in the year. Went at night to B. and F. I. soirée—a most brilliant affair. Moscheles there, and heaven knows all who besides in the musical line. '*22nd.*—At work as usual. Dined, came home, slept, dressed, took a cab to the Athenæum; met Sir Henry, and went with him to a soirée at the Marquis of Northampton's. Duke of Cambridge there, Lord Brougham, and many others; Hallam, Monckton Milnes, Forbes, Graham, Gifford, Babbage, etc. etc. '*2nd March,* Sunday.—Read and wrote. Walked through St. James's Park to Hyde Park, up Hyde Park along Oxford Street, and down Regent Street. Dinner, and came home to roast chestnuts, and finish the rough draft of a paper for our Memoirs.' This paper is again referred to under date 5th June, where the entry records : 'Writing at home at night. Finished my paper for the Memoirs, that is the first writing of it *sans* re-reading.' This was his famous essay on the 'Denudation of South Wales,' which eventually appeared in the first volume of the *Memoirs of the Geological Survey* in 1846, and of which some further account will be given in the following chapter. '*4th.*—Forbes's lecture. Dined with Falconer[1] at the Oriental Club; capital turn-out. Refused the Geological Survey of India. Heigho! Went to the Linnæan, and afterwards Forbes, Ibbetson, Henfrey, and I supped at Lankester's.' Further reference is

[1] Hugh Falconer, born 1808, died 1865 ; distinguished as a palæontologist and botanist, especially in regard to India, where he spent a large part of his life. His great memoirs on the fossil vertebrates of the Sivalik Hills have been of the highest importance in the history of palæontological discovery.

made two days later to the host here mentioned.
'Falconer came, and Logan[1] is the man for India.
Sir H. told me, like a daddy, he would advise me not
to go, but he would not stand in my way. Shan't take
it. At Captain Smyth's[2] at night. Pleasant party.
' 12th.—Reeks[3] came home with me, and we had
tea and ham together. Then the Geological Society;
scrimmage between Sedgwick and Greenough. Playfair and I had a long talk after about my Welsh affairs.'
The animated discussions at this Society are merely
alluded to in the diary. Thus on 5th February he
notes : ' Geological night. Fitton[4] on Greensand;
a tremendous row, and a regular blow-up after between
Fitton and Forbes.' On 2nd April: 'Went to the
Geological Society, where old Warburton[5] frightened
me out of my wits by calling on me to speak;'[6] and
on the 16th of the same month : ' Jolly night at the
Geological. Buckland's glaciers smashed.'[7]

[1] William Edmond Logan, born 1798, died 1875; connected in early life
with the South Welsh coal-field, of which he mapped a large part, afterwards handing
over his maps to the Geological Survey, which published them; subsequently
appointed Director of the Geological Survey of Canada; one of the great
pioneers of pre-Cambrian geology. He was a life-long cherished friend of
Ramsay. He retired from the Canadian Survey in 1869, and afterwards settled
in this country, and died here.

[2] Captain, afterwards Admiral William Henry Smyth, born 1788, died
1865; distinguished for his great survey of the Mediterranean, for his numerous
contributions on nautical and astronomical subjects, for his acquirements in numis-
matics, and for his important services in founding the Royal Geographical Society.

[3] Trenham Reeks was appointed in 1839 to the Museum in Craig's Court,
and was Curator of the Museum of Practical Geology in Jermyn Street from its
inauguration until his death in 1879.

[4] William Henry Fitton, born 1780, died 1861; an able geologist, to whom
we are largely indebted for the stratigraphical arrangement of the Cretaceous rocks
of England. The paper read by him on the 5th February 1845 was 'On the
Atherfield Section of the Lower Greensand in the Isle of Wight.'

[5] Henry Warburton, born 1784, died 1858; President of the Geological
Society 1843 to 1845.

[6] The subject on which Ramsay was called on to speak was a paper by Captain
Bayfield, 'On the Junction of the Transition and Primary Rocks of Canada.'

[7] The writer of this curt record lived to be one of the foremost supporters of
the 'glaciers' which he here dismisses. The paper read at the Society was one by
A. F. Mackintosh, 'On the Supposed Evidences of the Former Existence of Glaciers
in North Wales,' controverting the conclusions previously published by Buckland.

CHAPTER III

THE GEOLOGICAL SURVEY UNDER THE OFFICE OF WORKS

On the 1st April 1845, the beginning of the Parliamentary financial year, the Geological Survey was formally taken over from the Master-General and Board of Ordnance, and was placed 'under the direction and supervision of the First Commissioner of Her Majesty's Woods, Forests, Land Revenues, Works, and Buildings.' The staff was partly reorganised and somewhat augmented. At the same time the geological mapping of Ireland, which had been partially done for some of the north-eastern counties by Captain Portlock[1] under the Ordnance department, was now definitely undertaken upon the same lines as those followed in the larger island. The Irish Survey was made to form part of an organisation which embraced the whole United Kingdom, and which now became 'The Geological Survey of Great Britain and Ireland.'

The chief appointments of the staff thus enlarged were arranged as follows : Sir Henry De la Beche had charge of the whole organisation, with the title of Director-General. The immediate supervision of the

---

[1] J. E. Portlock, born 1794, died 1864 ; best known to geologists for his excellent memoir on the 'Geology of Londonderry, Tyrone, and Fermanagh, with portions of the Adjacent Counties.' He was President of the Geological Society in 1856-58.

work in England and Wales (and afterwards in Scotland) was assigned to A. C. Ramsay as Local Director for Great Britain. The Irish branch was entrusted to the care of Captain James,[1] R.E. A palæontologist was appointed, and the office was filled by Edward Forbes. W. W. Smyth[2] became Mining Geologist, and Dr. (now Sir Joseph) Hooker was a year later made Botanist to the Geological Survey of the United Kingdom. Dr. Lyon Playfair (now Lord Playfair) was appointed Chemist, while Richard Phillips still remained in charge of the original laboratory of the Museum. Robert Hunt succeeded T. B. Jordan as Keeper of Mining Records.

The staff of geological surveyors under Ramsay, besides W. T. Aveline, Trevor E. James, D. H. Williams, and H. W. Bristow,[3] already members of the Ordnance Geological Survey, was now augmented by the appointment of W. H. Baily[4] and of A. R. C. Selwyn[5]—a name which will be frequently mentioned in the course of this Memoir. Besides these officers, the staff included a few assistants for special services. R. Gibbs,[6] one of the most admirable collectors the

[1] Captain (afterwards Sir) Henry James, born 1803, died 1877; resigned the Directorship of the Geological Survey of Ireland in June 1846, and was succeeded by Thomas Oldham. He afterwards held for many years the appointment of Director of the Ordnance Survey of the United Kingdom.

[2] Warington W. Smyth, born 1817, died 1890; son of Admiral W. H. Smyth; Lecturer on Mining and Mineralogy in the Royal School of Mines from 1851 to the time of his death. Knighted in 1887.

[3] Henry William Bristow, born 1817, died 1889; appointed to the Survey in 1842; became Director for England and Wales in 1872; an able and accurate surveyor of Secondary and Tertiary formations, and, from his genial and courteous manners, a great favourite among his colleagues.

[4] William Hellier Baily, born 1819, died 1888; transferred in 1856 to the Irish staff, where he acted as Palæontologist.

[5] Alfred R. C. Selwyn, after doing admirable work in the mapping of North Wales, resigned, in July 1852, to accept the charge of the Geological Survey of Victoria. On the resignation of Sir William Logan, he was appointed Director of the Geological Survey of Canada, an office which he still worthily fills.

[6] Richard Gibbs, a native of Gloucestershire, was first employed by De la Beche in running sections in the Mitcheldean district, and made himself so useful

WILLIAM TALBOT AVELINE

Survey ever possessed, had joined as far back as the summer of 1843. Charles R. Bone was employed as artist to draw fossils described by the palæontologist. With this augmentation of the staff, other additional duties were undertaken by the Survey. Of these perhaps the most important was the preparation and publication of Memoirs illustrative of various districts that had been mapped, and containing a discussion of subjects connected with general views of geology and its applications. The first volume of this series was soon planned. The Director-General undertook to contribute an essay ' On the Formation of the Rocks of South Wales and South-Western England.' Edward Forbes supplied his famous and classic paper ' On the connection between the distribution of the existing fauna and flora of the British Isles, and the geological changes which have affected their area, especially during the epoch of the Northern Drift.' Ramsay's contribution consisted of his essay ' On the Denudation of South Wales and the adjacent counties of England.' The volume containing these various papers appeared in 1846, and the preparation of the material occupied much of the time spent indoors in the previous year.

The general bearing of the scientific organisation planned by De la Beche upon the progress of geological investigation was well expressed by Leonard Horner[1] in his address as President of the Geological Society. 'With scarcely any exceptions,' he said, 'all geological inquiries have [hitherto] been the

---

that he was eventually attached to the staff of the Survey. A large part of the fossil collections in the Museum of Practical Geology was originally collected by him. His name will frequently occur in the subsequent pages of this Memoir. He retired from the service on a pension in 1872, and died in 1878.

[1] See notice of Horner on p. 122.

fruits of individual research.   But in the Geological
Survey of Great Britain there is a combination of
forces which we have never, in this country at least,
seen applied to the promotion of any one department
of science.   No department perhaps requires so many
different descriptions of force to be brought to bear
upon it.   The Ordnance Trigonometrical Survey led
the way by the preparation of that indispensable
requisite in geological inquiries, an accurate map on
a large scale.   For the more general (geological)
Survey, we have geologists of great practical experience, who have established a high reputation; and
when the structure of each region is to be worked out
in detail, the special knowledge of the mineralogist,
the chemist, the natural philosopher, the zoologist,
the comparative anatomist, the botanist, and the
palæontologist, will be brought to bear, as required,
by means of men of high authority in each branch,
and their labours will be illustrated by artists of great
skill, all attached to the Survey, forming together a
corps of scientific men, for the accomplishment of a
great work, not surpassed, I believe, by any similar
establishment in any other country.'[1]

By these new arrangements additional duties
and responsibility were thrown upon Ramsay.   The
Local Director was to have immediate supervision of
the field-work of the staff, which would necessitate
his frequent inspection of the surveys of his various
colleagues.   He was to see that the whole mapping
was conducted on uniform methods, to confer with
the officers on their difficulties, to bring the experience
gained in one district to bear upon the elucidation of
another, and thus to ensure the harmony and steady

---

[1] Anniversary address, *Quart. Journ. Geol. Soc.* vol. iii. (1847), p. 31.

progress of the field-work. To gain these important objects it would no longer be possible for him to spend his whole time in the field carrying out independent mapping on his own part. It would be needful to keep himself in touch with the progress of the mapping in every district, though he resolved from the first that he would still devote a good part of the working season to mapping by himself—an employment for which he was so admirably qualified, and in which he took such a keen pleasure.

But besides superintending the surveys in the field, the Director was charged with the task of seeing the maps prepared for the engraver, of arranging the lines of horizontal section, and of editing these sections before they were sent to be engraved. These indoor duties were sometimes exceedingly onerous, involving as they did much correspondence and frequent visits to the ground before all discrepancies, omissions, or mistakes were finally rectified. From this time forward letter-writing on official business claimed an ever-increasing share of Ramsay's time.

The most irksome part, however, in the routine of these duties was the supervision of the accounts of the staff. Incredible as it may now seem, each member of the corps was required to procure a receipt for all travelling expenses. Continual and vexatious were the disputes with railway-clerks, coach-proprietors, hotel-keepers, and others who refused to be at the trouble of granting receipts, or declined even to sign their names at the foot of official receipts already prepared for them. Moreover, each officer was further bound to furnish at the end of every quarter a detailed statement of his disbursements, with vouchers for his travelling fares and other payments.

All these documents required to be checked and made conformable to the regulations, and the operation sometimes took several days, even if it was not further prolonged by correspondence as to inaccuracies in the charges, or in the method of stating them. But the crowning vexation came after the whole accounts had been examined and passed. In those days it was officially required that before sending in his accounts the Director should appear before a magistrate, and swear to their accuracy. In a country place, as may easily be imagined, this regulation often led to great loss of time, as well as additional expense. It would sometimes happen that no qualified official was to be found within a distance of several miles from the Director's station. And now and then, when found, the worthy justice had some difficulty in comprehending the nature of the unusual request that was made to him.

Ramsay chronicles a number of instances of his experience of this serious infliction in country places. Thus in the beginning of the October quarter of 1847, while at Bishop's Castle, he records : ' Got the accounts sworn to before Squire O.—a jolly, gentlemanly red-faced man, who did not seem clearly to understand the difference between an affidavit and an oath. Accordingly, as the surest method, he made me kiss the book.' On another occasion, while stationed at Llanberis in 1849, he had the experience recorded in the following entry : ' Having received the amended accounts, started for Mr. Hughes' of St. Ann's, the magistrate, ten miles off or so. He was away at Llanfairfechan. No help for it but to walk to Bangor. Every magistrate in the town was away to Llanfairfechan, for it seems they are one and all *parsons*, the magistrates of Bangor, and there was a chapel to be consecrated there to-day. Took a car, and in despair drove to Penrhyn Castle. Colonel Pennant also gone to Llanfairfechan!' Next day he proceeded to Caernarvon, 'expecting Mr. Morgan to do my magistrate's business for me, but lo! he was gone to Bangor, and no other was to be found in the town. I was disgusted beyond measure. Then took steamer and crossed to Anglesey to the Rev. Wynne Williams of Menai fron, and as good luck would have it, he was at home. At last we had run a magistrate to earth after a two days' hunt. He was very civil and made us take a glass of wine.'

The duties of the Local Director for Great Britain were at that time confined to England and Wales, the field-work not being extended to Scotland for some nine years later. The Irish branch of the service was entirely excluded from his supervision, but Ramsay was kept fully aware of all that was going on in the sister island, not only by conference and correspondence with De la Beche, but also by frequent communications from the successive directors of the Irish Survey. His tact and good sense were often of service in smoothing difficulties which threatened to break up the discipline and effectiveness of the Irish staff. Having the confidence both of the Director-General and his subordinates, he was appealed to frankly by both, so that over and above the correspondence naturally entailed on him by his own proper duties, he frequently was involved in much letter-writing on the affairs of his colleagues in Ireland. We get a glimpse into the life of the Irish Survey in the following unpublished letter from Edward Forbes.

FETHARD, COUNTY WEXFORD,
14th September 1845.

DEAR RAMSAY—When I arrived in London from my Zetland voyage I found you were in Glasgow. Had I known it before, I might have given you a ten minutes' call on the way. I got your note at Oban. On arriving in town I found half a dozen orders from Sir Henry to be off to join him in Ireland; so after three days in London, I cut away to Waterford *via* Bristol. . . .

I am here in a little village near Hook Point, in the midst of Mountain Limestone fossils, examining their distribution—all very interesting. The Captain, a very nice fellow named Willson, who is of his staff, and that thorough Welshman, little J., peppery, uncomfortable, and marvellously stupid and uninformed (as I find on close quarters), are my companions. We make a very merry mess, however, and the Welsh squire's absurdities—for he is in misery in Ireland—make us laugh. Sir Henry was with us till two days ago, working like a trooper, and when not at work telling funny stories. In a few days I leave this and go with the Captain (who sports a ferocious pair of egg-brown moustaches) to look at the Pleistocene beds in Wexford. Thence I go with Sir Henry to Dublin, after which, route as yet undetermined. When in Zetland I got most important data respecting the history of the animals found fossil in the Pleistocene beds. This makes me very anxious to see the Irish, and I should like much to go with you to Moel Trefaen and thereabouts.

I have been talking to Sir Henry about Longman's book.[1] I don't see how the Devil it is to be done. One gets no time to do it. Unless it can be done as Survey work, and in Survey time, it seems to me to be quite out of the question, and if we find that it cannot so be done, it would be better to write a joint letter to Longman submitting such to be the case, and requesting to be freed from the engagement. As it is, it is an unpleasant fiction. What say you? I have not finished my great work yet from utter want of time, nor when I think over it, do I see how it can be done, unless Sir Henry grants a few weeks' respite. Ever, dear Ramsay, most sincerely yours,   EDWARD FORBES △.

With all the immediate and prospective additions to his duties arising from the reorganisation of the Survey, and happily ignorant as yet of the trouble and worry which they might involve, Ramsay got through the office work of the winter of 1844-45, and soon

---

[1] This was the new edition of Conybeare and Phillips's *Geology of England and Wales*, referred to on p. 58.

after the middle of April took again to the field. He first joined H. W. Bristow in the Ludlow and New Radnor district, 'turning Old Red Sandstone into Silurian' during the day, and spending the evenings right merrily with musical friends.

The great Leopold von Buch, one of the oddest and ablest of the German geologists of his day, came to London early this summer. On the 27th May Ramsay notes: 'Adjourned to Dr. Fitton's, where were all the big-wigs of science to meet Von Buch.' A few weeks later he accompanied the German philosopher to Cambridge to attend the meeting of the British Association, which was held there in June of this year. Of this journey down from London he afterwards wrote: 'At Murchison's request I took Von Buch to Cambridge on the outside of the mail-coach from the head of the Haymarket. His luggage always consisted only of a small baize bag, which held a clean shirt and clean silk stockings. He wore knee-breeches and shoes.'[1]

At this meeting Ramsay read a paper before Section C 'On the Denudation of South Wales and the adjacent Counties.' His jotting in reference to this event runs thus: 'Read my paper, or rather spoke it. Felt no difficulty. Much discussion. Dined with the Reds. Evening meeting in our hall' [Jesus College]. Edward Forbes formed one of the merry party that was lodged in the College.

It was past the middle of July before he had resumed his field-work in Cardiganshire, working along the coast and into the interior from Aberporth, Aberaeron, Aberystwith, and other stations. This was to be a season of hard work in the field, clearing off

---

[1] *Life of Sir R. I. Murchison*, vol. ii. p. 76, *footnote*.

outstanding unfinished tracts of ground, and joining up lines so as to carry the mapping well to the north. It was only interrupted by a few weeks spent in a visit to his mother and friends in Scotland, during which he passed an evening on the top of Goatfell, renewed his acquaintance with the Glasgow circle, and saw his relatives in Edinburgh and Haddington.

Back again in Wales, he writes to his mother from Aberaeron on the 29th September : 'Stress of weather has delayed my work, two successive bad days having driven me to the verge of despair, and, had I good opportunity, there is no saying but I might run away to sea. I have been wandering after dinner on the shingly sea-shore. The wind was low, but a heavy smooth swell played the dickens with the pebbles, rattling and rolling them, and grind, grind, grinding them into rounded surfaces as polished as a smooth teapot. Then such piles of watery clouds in the west, full of portentous caverns, through which the upward rays of the sun (himself deep down in the sea) shone with a strange unearthly light, the whilk it was diffi-cult to say whether it most resembled a reflected glow from the gates of Heaven or a lurid glare from the portals of Infernality ! '

The Welsh ground that had to be mapped at this time included tracts that lay far from his stations, and necessitated long tramps on foot. Writing to his brother William from Aberystwith on 25th Sep-tember, he asks, ' Will no Christian make me a present of a thousand pounds ? and then I might buy a horse and gig and save my bones. When a man is wearied his brain is barren. That's my case. I could sleep, too, if it weren't that the tea keeps me wakeful. . . . I wish I had four legs and a man's head. I wish I were a centaur, and then I could go right across the country, taking all the hedges and ditches just as they come.' On 6th October he tells the same correspondent, ' I have been obligated to buy a pony, for this is too wild a country and the distances too great for my legs to stand it. The day before I reached Aberaeron I was fairly knocked up long before I reached home. Ten miles to one's work is rather too much of a good joke, for it makes twenty without including the work at all. I have got a great bargain, having only paid £7 : 10s. for her. She is at present well worth £12 or £13, and in six months I shall make her worth more than double what I paid. She is a chestnut, with silver mane and tail, and five years old last May.' Four days later he writes to his brother : ' I get a deuce of a drenching every day just now, even to the very sark. However, it does me no harm. My new pony turns out well—a little skittish sometimes, but that makes one feel alive in the saddle.'

The short November days would sometimes close in upon him while still far from his quarters, as on one occasion, of which he notes, ' Walked up the road to Llanidloes, and so over the shoulder of Plynlimmon. Benighted on the hills, *sans* road, and so dark I could not see two yards. By dint of shouting, a man came and found me.' At last, on the 14th November, he is able to chronicle at Pont-rhyd-fendigaid : ' Had a most successful day's work, and finished South Wales, perfectly understanding the same.' Before the end of the year he joined Selwyn at Machynlleth, and the two comrades made some traverses into the rugged country of Cader Idris, from which in later years they were to work out the complicated volcanic geology of North Wales.

While stationed at Pumpsaint, near Llandovery, in 1842, Ramsay had received much kindness from Mr. Johnes of Dolaucothi and his family. The friendship then begun was one of the most cherished of his life, and lasted undimmed to the time of his death. He was always a welcome guest at the house and a constant correspondent of the family. Not infrequently his epistles took the form of verse, and on his visits he sometimes wrote rhymes in the albums of the ladies. The earliest of these effusions dates back to the summer of 1842, and its character may be gathered from the following lines in it :—

> And when 'mid other scenes I ride,
> With good Sir Henry by my side,
> Oft will I tell of merry staves,
> Sung in Gogofau's ancient caves ;
> And how his 'geologic son'
> At Dolaucothi had 'such fun' ;
> Fenced with his host upon the green,
> And came off second best, I ween ;
> Ran races on the lawn, good lack !
> And tumbled down upon his back ;
> Or shouted loud among the train,
> Till woodland echoes rang again ;
> When I (with all the mirthful crew,
> Yourself, and B, C, F, and Stue),
> A stranger from the 'Land o' Cakes,'
> On Cothi's banks made ducks and drakes ;
> Or how, 'mid arbor-vitæ bowers,
> We plucked our ante-dinner flowers;
> And lofty Fanny chose to wear,
> Entwined amid her raven hair,
> Of cabbages a garland fair,
> While Charlotte, less ambitious, weaves
> A simple wreath of carrot leaves.

While the Geological Survey was in progress in Wales it was not difficult for him to pay an occasional visit to Dolaucothi, where he was always certain of a cordial welcome. And even after the field-work had been finished in the Principality he was able from time to time to return to this hospitable home.[1] From his voluminous correspondence with the Dolaucothi household we shall glean some interesting reminiscences of his life and work in later years.

From the brief entries in his memorandum book of 1846 a few quotations may be taken, giving glimpses into his London doings during three months in the early part of the year.

'21*st January*.—Went to Putney with Playfair. Lecture on chemical affinity. Came up to hear Sedgwick's paper on Wales, Cumberland, etc. Made a speech about South Wales. The old man horribly wrong-headed.' This meeting is referred to in a letter of the 31st January to W. T. Aveline : 'Sedgwick is at work attempting to show that we are all wrong, and that all North Wales (!), I think, and all *South Wales* —Cardigan and Caermarthenshire—is Upper Silurian.[2] He vows that Aberystwith is Ludlow. I flared up the other night, after his paper at the Geological, when he said that that was now the case, and thus we must not leave him the shadow of a leg to stand on. He is not content with the Cambrian, and so, gulping it down, he wheels about ten times, and turns it all in Upper Silurian.'

'29*th*.—At the Museum as usual. Had a scramble with Sir H. among the old book-shops after four. Bought an old Beaumont and Fletcher, and a Walton and Cotton. Evening at home. Wrote Eliza.

14*th February*.—At home at night reading the fifth edition of the *Vestiges* [*of Creation*]. Saw in it things

---

[1] Among his papers he preserved a clever and amusing sketch of a road map of Wales by Edward Forbes, which showed Dolaucothi in the centre, with roads leading directly to it from every quarter, even the most remote, where Ramsay was stationed.

[2] Yet Sedgwick was partly right. See W. Keeping, *Quart. Journ. Geol. Soc.* xxxvii. (1881), p. 141.

I had told Chambers in Edinburgh after the publication of the fourth edition. He *is* the author [see p. 137].

20th.—Anniversary of the Geological Society. Heard some of it. Went to the dinner afterwards ; sat beside Sir H. and Henry,[1] Ansted,[2] Strickland, Sopwith,[3] Austen,[4] and others of our party. Good fun. Murchison awfully grand.

21st.—At work as usual. Lord Northampton's first soirée. Prince Albert, Sir Robert [Peel], and Lord John [Russell] there among the crowd. Left at half-past eleven. Dined with Reeks and Baily.

25th.—Went to hear paper [by J. Prestwich[5]] on the Tertiaries of the Isle of Wight at the Geological ; excellent. The —— made an ass of himself. Sir H. spoke admirably.

12th March.—Sir H. criticised the first part of my paper [on the Denudation of South Wales] to-day most flatteringly.

17th.—Dr. Smith dined with us. Afterwards we went to the Haymarket, and died of laughter. Oysters with Playfair.'

[1] Thomas Hetherington Henry, at one time head brewer in the brewery of Messrs. Trueman, Hanbury, and Buxton, afterwards practised as an analytical chemist. He was a contemporary and friend of Edward Forbes, a member of the Red Lion Club, and was elected a Fellow of the Royal Society.

[2] David Thomas Ansted, born 1814, died 1880 ; Professor of Geology at King's College, London ; Assistant-Secretary of the Geological Society from 1844 to 1847 ; author of numerous popular works on geology. During the last thirty years of his life he was largely consulted in regard to the practical applications of geology, and was much employed as a professional witness.

[3] Thomas Sopwith, born 1803, died 1879 ; an ingenious mechanician, who devised many excellent geological models ; devoted much time to the study of mining districts ; became Commissioner for the Crown under the Dean-Forest Mining Act, and ultimately manager of the Allendale Mines.

[4] R. A. C. Godwin-Austen, born 1808, died 1884 ; a geologist of the keenest insight, who, though he wrote little, was acknowledged by his contemporaries to be one of the greatest of their number. His paper on the Possible Existence of Coal under the South-East of England is a remarkable example of his skill. He specially delighted in reconstructing the geography of former geological periods.

[5] Joseph Prestwich, the living Nestor of English geologists, specially distinguished for his researches in Tertiary and Post-Tertiary geology, succeeded Phillips in 1874 as Professor of Geology at Oxford, and retired from that office in 1888.

Ramsay remained in London until the 25th April, and passed the whole of the rest of the year in the field, partly inspecting work already done, and partly joining several of his colleagues in the tracing of fresh lines and in the attack of new ground. The staff had been recruited at the beginning of that month by the addition of another palæontologist in the person of J. W. Salter,[1] who was to play a brilliant part in working out the fossils of the Cambrian and Silurian formations of Britain, and who from time to time gave his services to the working parties in the field.

After a brief period of inspection with Bristow around Yeovil, in Somerset, Ramsay turned his face once more to North Wales, and remained there until the middle of December. The problems which the Survey was attacking in that region were of absorbing interest. Besides the question of the relative boundaries of the Cambrian and Silurian systems, then beginning to be agitated between Murchison and Sedgwick, there was the marvellous display of volcanic phenomena presented by the older Palæozoic rocks. Ramsay had already helped to map the igneous masses north of Builth, in the Shelve and Chirbury district, and among the Breidden Hills. But it was a new experience to see volcanic sheets developed on the magnificent scale which they present in the noble range of mountains extending from the ridge of Cader Idris northwards through the chain of the Arans and the Arenigs. Selwyn was at work from Dolgelli. Aveline was tracing the boundaries of the Silurian series from Llanbrynmair eastwards to

---

[1] John William Salter, born 1820, died 1869; appointed to the Geological Survey in 1846, and for many years engaged both in the Museum and in the field in determining fossils for the Survey. His knowledge of Palæozoic forms of life was unrivalled. He retired from the Survey in 1863.

Church Stretton and the Longmynd. At the beginning of October J. B. Jukes (see p. 82), joined the staff, and began to work out the ground around Bala. Between these three centres of field-work Ramsay spent some busy months, keeping himself in continual touch with the progress of the mapping, and taking also an active part in personally tracing geological lines. A few excerpts from his memorandum book will bring his life at this time before the reader.

'*14th May.*—Out with Selwyn along the front of the Cader cliffs. Glorious day and glorious scenery.

16th.—Out with Selwyn over the top of Cader Idris ; a long day's work. We had a splendid scramble.

20th.—Wrote Sir H. and Smyth with traps ; also Gibbs, Playfair, and my mother. Out by Ty-gwyn and all that country. Selwyn's work good. We wandered all day by mighty pleasant brooks and rivers.

1st June.—Left for Mallwyd ; met Aveline there.

2nd.—Aveline and I began to work. Work excellent, so far as I saw, especially the traps ; awfully hot day.

9th.—Out seeing the unconformable Caradocs on the Longmynd ; splendid old coast. Never more charmed.

23rd.—Wrote Eliza, and Survey letters to Trimmer,[1] Henfrey, Selwyn, Hunt. Walked up to Abergwailas with Aveline, working all the way. Dined there. Started again after dinner, and wrought till nine at night ; then back to tea.

---

[1] Joshua Trimmer, born 1795, died 1857. He was the first to recognise the importance of mapping the drifts and other superficial formations. He was for some time attached to the Geological Survey.

ALFRED R. C. SELWYN

24th.—No sleep at Abergwailas owing to green tea and fleas. Everything dirty. Started at ten, Aveline and I taking different routes. We met at five, having each traced some ten miles of winding boundary and meeting to a nicety. We got home to Pen y bont at nine, having walked thirty-five miles and fasted twelve hours. Mutton chops and bottled porter.

August 2nd.—Left [Dolaucothi] after lunch. Got awfully drenched on Lampeter Mountain; stayed at Lampeter two hours. Rode to Aberaeron; flood so great I was obliged to stay two hours there. Rode on; bridges all gone; forded the foaming torrents with difficulty; so late when I got to Llanrhystyd that I stayed at Lewis.

3rd.—Rode on at half-past six to Aberystwith. Breakfasted at the 'Lion.' Called and spent half the day with Fosset and the Downies. Rode on to Dolgelli, and got there at half-past twelve at night. Slept at the inn.

19th.—Out by Llyn bach and along the range of hills to the Dinas road. Splendid day.

25th.—Out on Rhobell fawr.[1] Excellent day's work.

26th.—Out above Llanfachreth. Home to dinner at six. Found Sir Henry dining in the coffee-room. Dined with the cricketers, and had a chat with Sir H. after.

27th.—Out with Sir H. over Cader; got many wrinkles. Met a car at Tal-y-llyn; home to dinner at seven. Selwyn and I dined with Sir Henry. Sir H.

[1] This interesting but difficult piece of geology was the subject of much careful exploration in later years both by Ramsay and Selwyn. When the survey of North Wales was almost completed, Ramsay wrote to Aveline (28th June 1853), 'I have often been prone to consider Rhobell as probably one of the centres of eruption.'

G

breakfasted with us.'   With reference to the 'wrinkles'
which the Director-General was able to furnish to his
younger colleagues, the following extract from a letter
written four days afterwards to W. T. Aveline is not
without geological interest: ' There are a number of
bands of strata here, which I at first took for altered
rocks, but which Sir Henry declared to be volcanic
ash, and which, though doubtless often deposited in
water, he declares must be mapped in green.[1]   His
reason is that they are volcanic products; and I
see he is right.   Some of them in structure are as
fine as porcelain - slates, being mostly or entirely
composed of felspar.'

' 29*th*.—Car to Drws y nant.   Over Aran Benllyn
and Aran Mowddwy to Dinas Mowddwy.   Came
home in a car to Dolgelli, well tired; Sir H. the
freshest of us.   Dined, slept, awoke, and went to bed.

4*th September*.—Started.   Walked down to the
castle [Harlech] after breakfast.   Splendid ruin.   From
thence along the coast, where we bathed.   Thence to
Llanbedr, and across the sandstone hills through Bwlch
Ardudwy to Pont dol gefeiliau, where we met a car
and drove home by eight.   Among the hills Sir H. and
I had lunch in a hut, where we met a Welshwoman,
who gave us bread, butter, and milk, and a hearty
welcome.

22*nd*.—Went with Selwyn, Oldham [see p. 84], and
Salter to Barmouth, etc. ; walked back across the sandstone hills over Llawllech.   Jolly day.   O. dined with us.

5*th October*.—Came up from Bala in a car with
Jukes[2] [who arrived at Bala on the 3rd to join the

---

[1] Green was the colour adopted at that time on the published maps of the
Geological Survey to express 'greenstones' and other igneous masses.
[2] Joseph Beete Jukes, born 1811, died 1869; after carrying on geological
explorations in Newfoundland and Torres Straits, returned to England and

staff of the Survey] to Ffestiniog, and found Sir H. there with erysipelas in the leg. Jolly!

10th.—Came down to Bala with Sir H. and Jukes in an eternal rain. Jukes and I dined with Sir H. Pleasant evening, very.

27th.—Jukes and Forbes rode to the foot of the Arenig; Aveline, Williams, Gibbs, and I walked. Foggy on the top. Ash, ash, ash everywhere.' By this time Ramsay's eyes were fully opened to the great importance of recognising the detrital material of old volcanic explosions among the Palæozoic systems. Writing to W. T. Aveline on the 23rd September, he says : ' On the whole, my experience here makes me much more sceptical of altered rocks, generally speaking, than I used to be, there being many beds here that I would once have considered altered rocks, which are in reality nothing but hard consolidated ashes. The word "ashes" does not imply "cinders," but often rather volcanic *dust*, which may be as fine as you like.'

' *18th November.*—Bala. Out among the traps ; had a glorious find of fossils.

19th.—Wrote Forbes with a trilobite. Away up far among the traps. Got an excellent day's work done.

21st.—Out with Aveline to the traps. It got wet and turned us just when we had begun to work. Nevertheless, we got a goodish day's work done on the way home by the river, into which, a branch giving way, I tumbled considerably over the boots.'

One of the most important events in the progress

joined the Geological Survey in 1846. He was appointed Director of the Geological Survey of Ireland in 1850, and held that post till his death. Much information about the Survey work in Wales will be found in Jukes's *Letters* (1871), many of which were written to A. C. Ramsay.

of the Survey, and also in the career of its Local
Director, during the year 1846 was the appearance of
the first volume of the Survey Memoirs already referred to.   Allusion to the coming volume was not
infrequent during the summer in the correspondence
of the Survey officials.   Thus Sir Henry, who had
gone over to Ireland to inspect the field-work there
under the new Director, T. Oldham,[1] wrote :

NEWTOWN BRAY, CO. WEXFORD,
26th July 1846.

MY DEAR RAMSAY — Oldham and self continue to get on
famously, and I am right well contented with him. . . . Oldham
appears to have a philosophical mind, quite ready to go ahead
in the school we have been forming.   In about ten or twelve days
I hope to be on the start for the other (your) side of the Irish Sea,
running up to London to see what progress we are making towards
a house.

Our tome of Memoirs is described as a handsome one; I
believe it to be a good one.   The Longmans say it is too *cheap*,
but somehow £1 : 1s. seems a fair price for any work of the kind.
—Ever yours,   H. T. DE LA BECHE.

We are here concerned with only Ramsay's contribution to the volume.   His essay on the Denudation of South Wales was a remarkably original and
suggestive addition to the literature of geology.   It
was the first attempt to reduce the phenomena of
denudation to actual measurement by constructing
horizontal sections on a true scale, and showing what
thickness of rock had actually been stripped off the
face of the country.

The following correspondence will show how this
essay was regarded by two of the ablest reasoners in
geological science :—

[1] Thomas Oldham, born 1816, died 1878; appointed to the directorship
of the Geological Survey of Ireland in 1846 on the retirement of Captain James,
and held it until 1850, when he became Superintendent of the Geological Survey
of India.   He retired from that office in 1876.

C. DARWIN TO A. C. RAMSAY.

DOWN, FARNBOROUGH,
10th October (1846).

DEAR SIR—Having just read your excellent Memoir on Denudation, I have taken the liberty to send you a copy of my volume on South America, finding that we have discussed some related questions. I wish I had profited by your Memoir before publishing my volume. I see that we entirely agree on the sea's great power compared with ordinary alluvial action, and likewise on the frequency of grand oscillations of level, and on several other points. If you had time to read parts of my volume, I should much like to discuss with you many cases, such as my notion of subsidence being necessary for the formation of high sea-cliffs, as *inferred* from the *nature of the sea's bottom off them;* likewise the horizontal elevation of the Cordillera as inferred from the sloping gravel fringes in the valleys ; the non-horizontality of lines of escarpments round old bays, etc. etc. I grieve to see how diametrically opposite our views are (I being a follower of Lyell) on the probability of great and sudden elevations of mountain-chains ; I cannot but think that you would have estimated existing forces as more than 'petty,' and entertained some doubt about their being 'conflicting,' had you inspected with your own eyes the wide area of recently elevated and similarly affected districts in South America. There is much which I could say on this head, but I will not intrude on you.

May I ask whether you do not admit Mr. Hopkins's views of mountain-chains being the subordinate effects of fractures consequent on changes of level in the surrounding areas ; and does not all the evidence which we possess tend to show that widely-extended elevations are slow, and may we not infer from this that the formation of mountain-chains is likewise probably slow ? I cannot see any difficulty, after a line of fracture has been once formed, in fluidified rock being pumped in by as many strokes, as it is pumped out in a common volcano, and yet producing a symmetrical effect. But I much fear that I have cause to apologise for having written at such unreasonable length ; the interest excited in me by your Memoir must plead my excuse, and trusting that you will forgive the liberty I have taken, I remain, dear sir, yours faithfully,   C. DARWIN.

C. LYELL TO A. C. RAMSAY.

MY DEAR SIR—I have just been reading with great pleasure your admirable and well-written essay on the Denudation of South

Wales, the illustrations of which are most beautiful, and with nine-tenths of which, in regard to the conclusion and reasonings, I agree. I shall have to cite it in a seventh edition of my *Principles*, which I am now printing, and I wish to guard myself against misunderstanding the only point on which I shall have to differ from you. I am anxious to have a speedy answer, as the sheet in which I shall allude to your paper will soon go to press.

In that part of your section, plate 4, which relates to the Mendip region, one which I have gone over sufficiently to take more interest in it than I otherwise should do, you mean to say, if I mistake not, that between 4000 and 5000 feet of strata have been removed by denudation, between C and 4*a*; in other words, that between the deposition of certain Carboniferous strata and certain newer beds, 4*a*, two events occurred: first, the disturbance of the beds of the Palæozoic rocks; second, the denudation of several thousand feet of the same beds. Yet it seems to me, on reading other passages of your paper, that you cannot mean this; for your ideas of denudation acting contemporaneously with subterranean movements, whether of upheaval or depression, agree with those which I published in 1831 in my *Principles* (and, by the way, before that time it was thought a triumphant argument against what were called 'modern causes' to prove that a river could not denude the rocks); the gradual action of the ocean acting concurrently with movements of the land, as exemplified in my denudation of the Wealden, had not, so far as I know, been fully set forth in any geological work, with due allowance also for the resistance of the harder and yielding of the softer rocks.

Now as you adopt these views, and have applied them with all the modern lights to your sound and philosophical speculations in this essay, I cannot comprehend how you can dispense with indefinite geological time for your denudation in the case I allude to. But if so, what becomes of your argument at p. 317 in favour of grand catastrophic and intense disturbing power between the close of the Coal Measure Period and the deposition of the unconformable beds, 4*a*? You must be well prepared, from what I have said of the amount of denudation, especially in my *Elements*, for my willingness to admit as much of it, or more than you want, as it must have exceeded all the sedimentary strata which are a measure of its quantity, and of the gradual manner or slow rate at which it took place. You will also expect that I, at least, shall feel no difficulty in granting an indefinite lapse of geological time between the deposition of the last of the Coal Measures in the Mendip region and the oldest of the overlying conglomerates of that country. The entire flora and fauna were changed once, if not more, during that interval of unknown extent, and therefore I have no objection to as much elevation, disturbance, and denudation as you represent. But as I interpret your paper, you think the period so short that you have only time for one gigantic effort to cause all the faults, fissures, and curvatures of the older strata. You may say, perhaps, that the greater time taken for the denudation the less remains for the upheaval; but that argument is then against the two operations having been contemporaneous. That you do not suppose them to have gone on simultaneously I conclude from the suddenness which you attribute to the action of the disturbing forces. You also seem to assume (p. 317) that the action of forces working at successive times could be conflicting. This seems to be so contrary to the analogy of volcanic action, whether breaking out at the surface or exhibited in the upheaval or depression of continents, or in convulsions which rend for several thousand years lines of country, recurring so marvellously in the same tracts, and in the same direction, that some facts or some references to contrary analogies should have been given. The conformity of the Palæozoic strata can in no wise circumscribe the lapse of time of which I have defined the limits as above.

I do not think that any geologist who has lived, as we have done, in a period when a single earthquake can rend a large district like Chili, and permanently uplift a portion of the earth's crust, which may possibly be miles in depth, will quarrel with you for any intensity which you ascribe to the disturbing power; but I shall be surprised if you do not live to see the day when few will think it consistent either with the ancient Plutonic or Trappean phenomena, or with our acquaintance with actual igneous action, to suppose that so mighty a change in the interior of the earth occurred at once, as is implied by the sudden uplifting and contortion of thousands of feet of strata. That no relief should have been obtained by intermittent action, as now by the rending of the crust as soon as the expansive power of the melted matter requires more room, but that it should have all been kept under till it could be accommodated during one grand convulsion is, I suspect, an hypothesis unnecessary on mechanical grounds, and especially undesirable by one who adopts *our* views of denudation, which are so naturally aided by taking not only unlimited time for the development of igneous action, but equally so as regards the upward and downward movements, and the rending and bending of the beds. But I do not write this to make a convert of you, but that you may explain if I have misconstrued your meaning, and you will better see how I interpret you by my entering into this line of objection. I do not quite follow you on the argument founded on the missing members; but I am sure you cannot assume that, in a region suffering denudation, deposits of such a nature as to last must be found in the immediate neighbourhood. It is, I believe, quite the exception to the rule, and in this view I am not singular. Hoping to hear from you soon, and with many thanks for one of the very best papers I ever read, believe me, very truly yours,   CHA. LYELL.

11 HARLEY STREET, 8th October 1846.

When I inferred that the denuded dome of the Wealden had lost some 2000 [feet] and upwards of thickness of strata removed, I also assumed that it was shaved off by the ocean when rising, and had never constituted hills 4000 to 5000 feet high. So I think of your denuded tracts. They were never suffered to attain an Alpine elevation.

To the foregoing letter Ramsay sent the following reply :—

BALA, 19th October 1846.

MY DEAR SIR—We have been so busy in the field (Sir Henry only having left us to-day) that I have not previously had time fully to consider your letter.

In the beginning you refer to what I have written about the Mendip Hills as a type of certain denudations in the following words : ' First, the disturbance of the beds of the Palæozoic rocks ; second, the denudation of several thousand feet of the same beds. Yet it seems to me, on reading other passages of your paper, that you cannot mean this ; for your ideas of denudation acting contemporaneously with subterranean move-ments, whether of upheaval or depression, agree with those which I published in 1831 in my *Principles;* the gradual action of the ocean acting concurrently with movements of the land, etc. etc. Now, as you adopt these views, I cannot comprehend how you can dis-pense with indefinite geological time for your denuda-tion in the case I allude to, etc. etc.'

In this argument I think you have partially mis-understood me, partly because, in my anxiety to be concise, I have not sufficiently entered into detail ; and again perhaps from not bearing in mind that my reasonings are not intended for universal generalisation, but simply refer to the phenomena exhibited in a given district which came during the progress of the Survey under my especial notice.

First, With regard to the disturbance and upheaval of rocks. You will observe that in different parts of this essay I recognise two distinct species of phenomena, one that of violent and extensive disturbance on a gigantic scale, and another that of slow upheaval and depression, such as we now have experience of on the shores of the Baltic and elsewhere. With regard to the latter, I grant that any possible height may be attained by an unlimited succession of comparatively small elevating disturbances. But in reference to the former I cannot conceive that the early forces that affected the Palæozoic strata in the district treated of, and which I believe then elevated these strata, were at all of the same intensity as those that in later periods occurred in the same tracts. You may raise to a given height a certain bar of iron with your finger, but no succession of the same forces, however numerous, could crush that bar laterally like a plaited frill. A more sudden and powerful effort is requisite. Such I believe to have been the case with the Mendip Hills, viewing them solely as part of a much wider area. But even granting this to be the case, it does not follow that I therefore restrict the period required for denudation. In the case referred to (the Mendips) you will observe that I state, p. 320, that 'a mass of limestone, etc., once existing above part of the Mendip Hills, to an extent of at least 6000 feet high, had been removed by the denuding agency *of the New Red sea,* or *possibly by that sea and the earliest Liassic waters,* since we find the lower beds of the Lias resting horizontally on the upturned water-worn edges of the

rocks that now form the flattened summit of the range.'
Denudation, I uphold, though materially affected by
the nature (suddenness and intensity) of disturbances,
yet as an independent agency acts in a measure independently of them.   Thus the sea has greater facilities
afforded it for acting on a coast subjected to frequent
and gradual submergence and emergence ; but the sea
would still act, and might, *if time were allowed it*,
utterly destroy a tract of land of any given height, with
little or no oscillation of level.   In the instance of the
Mendip Hills, there were doubtless many minor oscillations after the great catastrophe, and you will observe
from the foregoing quotation that I allow great part of
the New Red and part of the Liassic epochs for the
sea to effect this denudation.

This surely was a period of time fully equal in
extent to those comparatively latter days to which I
refer the greater part of the South Wales denudations.
I think in the paper I have given physical reasons to
show that in South Wales and the neighbourhood the
greatest denudation of *great part* of these tracts did
not take place during the first elevation (pp. 317, 323).

I shall by and by have occasion in subsequent
papers on the same subject to prove that in other
countries such long-continued denudations did take
place during a series of disturbances that affected the
Palæozoic strata during the very period of their formation.   The whole matter is merely a question of
degree, denudation in both cases being produced by
changes of level.

Respecting the forces that produced the earlier
great disturbances, I have perhaps used the words
'conflicting forces' somewhat hastily.   My object was
to show that the remarkable and intense curvatures

with which that country abounds were not produced
by any efforts of melted matter to escape, these efforts
being attended with the curved upheaval of different
parts of the strata where the efforts were made. I
wished to show that the curvatures were due to a more
universal action and a different power (see De la
Beche's *Theoretical Researches;* I have not a copy
by me), viz. an effort of contraction in a solid crust to
accommodate itself by the force of gravitation to a
cooling and lessening internal mass. I never dreamed
of doubting that relief was obtained from the very
beginning of geological time by intermittent outpour-
ings of melted matter. On the contrary, I have got a
quantity of data together for future use respecting
remarkable periods when visible volcanic action more
or less obtained, and the efforts produced by this
action on the strata of the day ; but I shall also be
able to show that all these volcanoes were subsequently
affected equally with their associated strata by a power
stronger than they or than that which produced them.
You must not suppose that by such efforts I understand a universal general crumpling at one time of
the whole earth's surface. There is perfect evidence
that the contrary was the case. It is not the *accommodation* of the melted matter I contend for, but the
accommodation of the solid circular external crust,
attempting to fit itself through the influence of gravitation to a cooling and diminishing area within.

Respecting the *missing members* of the New Red
Sandstone, I believe that they may exist, though
concealed by the Upper Series, it being remembered
that the New Red Sandstone lies as it were in a basin,
since we find that proceeding from north to south the
higher marls gradually encroach on and cover up the

lower beds. These lower beds may also cover up members still lower in the series so completely that they are nowhere exposed.

When I wrote this paper I was buried in the country without the means of reference to a single book, and if I have omitted to refer to what has been done by you and others in the same walk, it entirely arises from forgetfulness or ignorance. The papers referred to in notes were quoted from memory, and the notes in the manuscript left with blanks for more accurate detail when I got to town. Nothing could be further from my wish than to assume as my own any idea started by another, especially by one whose *Principles of Geology* strongly tended to make my geological mind such as it is by first directing its inquiries into proper channels, when, now some ten years ago, I first began to dip into geology as a relief from the irksome drudgery of mercantile concerns. Had I no higher motive than my affection for the mere book, that of itself would be sufficient to deter me from such an attempt.

Believe me, the opinion you express of my paper has been to me a source of no ordinary gratification, feeling as I do the value of approbation from so distinguished an author.—I remain, dear sir, ever yours sincerely,   ANDW. C. RAMSAY.

The following three letters of the same period afford an indication of the relation of the Geological Survey to the dispute regarding Cambrian and Silurian which was arising between Sedgwick and Murchison :—

MUSEUM, *29th December* 1846.

MY DEAR RAMSAY—I put you in possession of notes which have passed between Murchison and self touching Silurians, so that you may know how to treat things if any discussion arises at the next meeting of the G. S. The lower rocks should be mapped as we have proposed.

Sedgwick at last meeting spoke highly of our sections.

We have to keep a straight honest course, thinking only of truth, and aiding the advance of knowledge.

I am off in an hour or so for Swansea.—Ever yours,

H. T. DE LA BECHE.

NURSTED HOUSE, PETERSFIELD,
29th December 1846.

MY DEAR DE LA BECHE—Your note of the 24th followed me.

I cannot for a moment suppose after all you have *said* and *done*, and after your fair and public recognition of my 'Lower Silurian' types, that you can in any way intend to *swamp* them. The case is indeed so palpable that I believe every geologist is desirous of sustaining the names of the person who first worked out the succession from a known base line. All I expected, and do expect from you is—that *if* it be proven that Cambrian and Lower Silurian are geologically synonymous, you will adhere to my name— the only one worked out on a fossil basis. Now all I beg your permission to do in my little 'apology' for the Silurian System (the *Bible* has even been apologised for!) is that I may say 'by whatever name the rocks be defined, it must be *one name ;* for that the inquiries of the Government geologists (which are yet, however, not completed) go to prove that there is but one natural series or system of organic life in North Wales.'

Is there any objection to this, which leaves you by your *subsequent* inquiries to make any statement you please? At all events, the question between Sedgwick and myself is decided by his own evidence of the existence of some of the commonest Caradoc fossils in some of his very lowest beds.—Ever yours most sincerely,

ROD. I. MURCHISON.

GEOLOGICAL SURVEY OFFICE,
29th December 1846.

And now, my dear Murchison, a word for you before I start for South Wales, and that in about two hours. To tell you the truth, I was a bit inclined to look queer at the preamble of your last missive, seeing that you spoke a possible thought on my part of swamping your 'Lower Silurian types,' and this at the very time I had been doing my best to show how much I appreciated your labours, and so I did not say so much about the matter as I might. The said queerishness having duly evaporated, it is but fair that as you are going to write, and you have broached the matter to me, I should

put you in possession of things as I know them, since, as matters
stand, you will not get them elsewhere.

First to clear the ground as to classification.   If the older
rocks be classed, as they are, according to the remains of life found
in them, it follows that any given mass of them containing the same
kind of life, really and truly, should have but one name.   Whatever
the Bala beds may turn out to be, as to equivalent deposits in
geological time, they present, so Forbes says, the same *kind* of life
as that contained in the Silurian system, according to your published
works.   The consequence of general name follows—that is, one
name for the whole.

Next come the beds, which have been termed the Lingula beds
—these underlie the others.   Whatever other fossils may be found
among them in Wales, our collection will ultimately show.   In the
meantime, rocks in Ireland which *may* be equivalent both to these
and the Bala beds contain Silurian fossils.   Supposing further
researches to confirm these views—the same kind of life still, with
its consequences about one name.

Beneath these come rocks which you know well at the Longmynd, and of which you showed me an old section, at least one
showing your views at the time respecting them.

Without aid of any kind Ramsay this year made out their story.
Of the equivalents near St. David's you have often heard me speak,
and touching the Irish rocks of the same date you know what I said
the other night.   Of all the exhibitions of them the Irish is the best
—a great thickness.   Now it has been supposed that these beds are
not fossiliferous.   This is not true in Ireland.   Two or three years
since Oldham got some things that were clearly organic, though in
such a condition that nothing could be made of them.   But this
year one of our lynx-eyed collectors has been turning out good
specimens.   What they are is not clear ; anyhow they are more
diffused than at first thought.   Now with this group of rocks
classification is not clear.   We have abundant evidence, capital in
Ireland, of their slow deposit through a long lapse of time.   These
things I tell you in confidence, because the affair is incomplete, and
there is much yet to be done, but you should know them—and now
I must run for it.—Ever sincerely,   H. T. DE LA BECHE.

Griffith, in a document he has sent in to us of about a year's
date, has got there older rocks, though not correctly mapped.   He
calls them Cambrian, and whenever he publishes his new edition of
map, so they will, I suppose, be called.

Before the end of the season for field-work in the
year 1846 Ramsay received an invitation which

gratified him not a little, as it offered to him a wider opening into the society of men of science and their associates. He was asked to give a Friday evening discourse at the Royal Institution, and on communicating to De la Beche that he had accepted this offer, he received from his chief the following characteristic note in reply :—

[CONFIDENTIAL.]

LONDON, *24th November* 1846.

MY DEAR RAMSAY—In the matter of an evening at the Royal Institution you have, I think, decided well—though I have by no means the view of any great good to arise therefrom which some folk have. The thing is over-rated.

All appliances shall be at your command—all aid that can be given, and I have little doubt of your coming out of the matter in good style.

Take some good commanding subject. Continue to think me a kind of daddy—the more you do that the better I shall be pleased. —Ever sincerely,   H. T. DE LA BECHE.

All advancing famously.

The field-work of this year was prolonged by Ramsay until late in the season. About the middle of December snow fell thickly over the Welsh hills, making further field-work for the time impossible. He accordingly took advantage of the opportunity to spend Christmas and New Year at his old home in the north. He announced his advent to his mother in the following effusion :—

BALA, *12th December* 1846.

MY DEAR MOTHER—

Plunge the poker in the fire,
    Stir the blaze ;
Rouse it high, and rouse it higher,
Till your very eyes it daze,
    Brightly glowing,
And the puny candle's rays
    Pale are growing.

Hark the kettle's cheerful songs,
        Shrilly crying
To the dull recumbent tongs,
On the clean-swept hearthstone lying,
        Cheerily singing !
' Blithe I be because I'm trying,
        By my ringing,

' To announce a merry meeting ;
        I am humming
To arouse a jolly greeting.
Hear me on the hob a-bumming ;
        Let me threep o't.
Even now I hear him coming ;
        Fill the tea-pot !

' Fill the tea-pot !   By the fire
        While you're basking
You shall have your heart's desire.
I will tell you without asking
        One is hasting
Even while the tea is masking,
        No time wasting.'

Mother, mother !—yes, I'm hasting,
        Hasting home, too ;
Posting, steaming, no time wasting,
Till, O happy day ! I come to
        Your own ingle,
For a while no more to roam to
        Dell and dingle.

Yes, I'm coming !   Rouse the fire ;
        Make it roar there ;
Rouse it high, and rouse it higher ;
Soon you'll hear me to the door there
        Madly bounding,
Then rushing in athwart the floor there
        With laugh astounding.

It's a fact, though I did not know it till too late for
the post.   If the weather be good I shall remain here,
as in duty bound, till the 15th or 16th.   If the snow
continue I shall leave to-morrow and be home, I suppose, on Tuesday to dinner, for the steamer leaves

Liverpool at seven at night. Therefore air the sheets and clear out that chest of drawers for a fortnight, for that's my allowance. What's the use of writing more when we shall have a fortnight to haver? With hurrahs to all, yours affectionately,

ANDW. C. RAMSAY.

These pleasant glimpses of the old home had always in prospect the inevitable pang of parting at their close. He had to leave upon his return on the 2nd January. Under that date he records that his mother and sister 'kept up well by dint of talking of all sorts of things till starting time came, when all at once they gave way. I did so also *almost*. I wish their hearts were a trifle harder. It almost makes going home painful, the dread of the leaving day.'

H

CHAPTER IV

THE PROFESSORSHIP OF GEOLOGY AT UNIVERSITY
COLLEGE, LONDON

THE year 1847 was to be a memorable one in the life
of the subject of this biography. The early months
of the year were passed in the usual official and social
engagements, into all of which Ramsay entered with
much zest. Perhaps the most important event in his
London life during the season was his introduction
to the general scientific society of London on the 12th
March, when he gave the Friday evening discourse
at the Royal Institution referred to in the foregoing
chapter. He chose as his subject one which his
Survey work in Wales had now made familiar to
him, and on which he had much fresh information to
convey—'The Causes and Amount of Geological
Denudations.' In later years he not infrequently discoursed in the same theatre, and usually with some
trepidation beforehand. The success of his lecture
always depended upon the mood he happened to be
in at the time, and he never could tell how he was
succeeding until his task was half done. 'The
Royal Institution,' he once wrote to me, 'is the most
ticklish audience in Britain to lecture before, because
the most critical and refined, and possessing also, in
large and equal shares, so much knowledge and so

HENRY WILLIAM BRISTOW

much ignorance.' We shall find in the course of this biography that the impression made by the young lecturer's exposition upon the present occasion had its influence in securing for him the goodwill of the authorities at University College.

By the beginning of April he had finished all the indoor work of the winter, and was ready to take the field. From that time, with the exception of the brief interval required to attend the meeting of the British Association at Oxford, where he acted as Secretary of Section C, he remained all the summer and autumn in the field.

A few weeks of inspection duty were first spent with H. W. Bristow among the Jurassic rocks of Dorsetshire, and a visit was paid to Aveline and his family near Wrington, in Somerset. By the first week in May he was once more back in Wales, running boundary-lines that still needed completion to the north of Brecon, and revising the volcanic geology to the north of Builth. His growing experience of ancient volcanic rocks now enabled him to separate the 'ashes' from the ordinary sediments, which in the earlier surveys had been grouped together, and to introduce much more precision and detail into the mapping of these rocks. W. W. Smyth, who had previously mapped part of the district, joined him at Builth, and the two colleagues re-examined the district together. On the 21st May Ramsay chronicles: 'Out with Smyth over the Carneddau, and on the traps farther north. Found them so egregiously wrong, that they will not stand an hour's investigation in the new style of mapping. Great part of the work to-day was revising his old work, so that we are all more or less in the same mess. The day

was lovely, and a splendid full-flavoured Havannah
we did enjoy, where the two brooks meet to the
east of Maes gwyn, reclining on the grass in the
sun with all the trees growing green under our very
eyes. That *was* a smoke. Continued the lines from
thence all down to Tan y graig, and got home to
two cold legs of lamb at six; Smyth finished the one,
and I the other. Wrote letters at night.' The com-
missariat of these Survey parties was sometimes apt
to vary a good deal both in quantity and in quality.
A few days after the disappearance of the two legs of
lamb, Aveline joined the party at Builth, bringing with
him a pig's cheek as a contribution to the common
larder. Ramsay records the consternation and distress of their worthy landlady on discovering that a
dog had got into the house, broken the dish, and made
off with the pig's cheek.

   ' *3rd June.*—Out with Aveline at the traps to the
north to show him my system of working them, that
there might be a unity in the work of the Survey.
Got some capital work done to the north of Llwyn
Madock.

   ' *4th.*—Wrote my mother on hearing of the death
of Dr. Chalmers.[1] He died suddenly in his bed, of
apoplexy, through the night on Sunday, the 30th. He
was the greatest and the best man I ever knew. I
am glad I saw him twice within the last two years,
and shall often think of his cheerfulness at the family
breakfasts, when, after the meal, his daughters, he, and
I used to " form the segment of a circle round the fire,"

   [1] Thomas Chalmers, born 1780, died 1847 ; the most distinguished preacher
of his day in the Church of Scotland. He was for some years minister of the
Tron parish in Glasgow, became Professor of Theology in the University of
Edinburgh from 1828 to 1843, and was the leading spirit in the secession that
founded the Free Church, of which he was the first Moderator. He was on
friendly terms with Mrs. Ramsay and her family.

when he described his Sunday congregation in the Cowgate "sitting in all the glory of their rags," and when, speaking of the Survey, he said, "I hope they endow you liberally, and do not engraft upon you the scurvy economics of a Joseph Hume."

'11th.—Heard from Sir Henry asking me if I would accept the Chair of Geology at University College, London, if it were offered me. Wrote favourably, but said I would not accept if the trouble were to be infinite and the emoluments nothing but the honour. I would accept on the grounds that the position of professor in one of the chartered Universities is a good addition to my present honourable position; that it would do me good in London scientific society, besides that a tag is a useful thing to a man's name anywhere. This is the second time it has been offered; the last was before Joyce was appointed.'

Two days afterwards he heard that the number of students attending the class of geology at University College had, during the previous five years, ranged from eight to seventeen, and the emoluments from £14 : 10s. to £48. The prospect which these figures opened out to him could not be regarded as inviting. Nevertheless he determined to accept the appointment if the season and hour of lecture could be made to fit in with his duties in the Survey. Edward Forbes wrote to him urging him to make his conditions definite, and particularly that the lectures should be given in one continued course in January, February, and March, and he added that 'remuneration can hardly be expected from it, but the position and title are worth having, and might under various circumstances be of much importance to you. I think, too, you might make a class in the end, and make it worth while. Such a continuous course of lectures would, at any rate, be excellent exercise. I fancy the disposers of the chair and the advisers of the Senate on such an occasion would be Greenough, Warburton, Hutton, and perhaps Fitton, so that you might be sure of just judges at least.'

As a mere matter of form he was asked to send in an application for the appointment, together with any testimonials he might wish to submit. Among his papers he has preserved the recommendation he received from Lyell, with the memorandum written on it, ' Not used, appointment being completed before it came.' The letter is as follows :—

KINNORDY, KIRRIEMUIR, N.B.,
7th July 1847.

DEAR SIR—My residence at a distance from London will explain the delay in my answer. I am glad to learn that you are a candidate for the Chair of Geology now vacant in University College, especially as I hear there are from sixty to seventy students at King's College. Several weeks ago one of the U. Coll. Council asked me my opinion in regard to your qualifications for such an office, and I mentioned your publications, so far as I knew them, from that on *The Geology of Arran* (the excellence of which I could test by my own knowledge of that island) to your last elaborate paper on ' Denudation ' in the *Memoirs of the Survey of Great Britain.*

I also alluded to your experience in field-work and the application of geology to the arts. I was naturally questioned on your powers as a lecturer, of which I was able to speak as having heard you at the Royal Institution. I said you were able to express yourself freely on subjects with which you were familiar without reference to MS., and that I thought with practice you would make a good oral instructor.

You may make any use you please of this letter, and I shall be happy to answer any further queries if required.—Believe me, ever truly yours,   CHA. LYELL.
A. C. Ramsay, Esq.

De la Beche warmly supported the claims of his colleague, and readily arranged that the duties of the professorship and those of Local Director of the

Geological Survey should not interfere with each other. It was agreed that the lectures should be given during the first three months of the year, and in the afternoon, so that he could complete his work at Craig's Court before going up to University College. In due course he was appointed to the chair, and now became 'Professor Ramsay,' the name by which he is best known, and which he continued to use for thirty-four years, until knighthood was conferred upon him at the end of his official career.

The British Association assembled this year for the second time at Oxford. Ramsay made a hurried journey to the meeting, and returned to his field-work in Wales. To him one of the pleasantest features of the week was the presence of his old friend Professor Nichol of Glasgow, whose early kindness he was enabled in some measure to repay by introducing him to a number of men of science whom the astronomer had not before met. From his notes of the meeting a few quotations may here be inserted.

' 24*th* June.—At eleven the business began with Chambers's paper on Raised Beaches. He certainly pushed his conclusions to a most unwarrantable length, and got roughly handled on account of it by Buckland, De la Beche, Sedgwick, Murchison, and Lyell. The last told me afterwards that he did so purposely that C. might see that reasonings in the style of the author of the *Vestiges* would not be tolerated among scientific men.

' 28*th*.—Prince Albert, the Crown Prince of Saxe-Weimar, came to our Section while Count Rosen was reading a paper, Murchison in the chair. This delighted Sir Roderick. Afterwards Dr. Buckland took the chair. Among others I read a paper on the

Contour - lines of Cardiganshire, showing that the hill-tops in the west formed a regular inclined plane to the sea. I did not read, but spoke it. It was well received by all but the Dean.

'*29th.*—Sir Henry was in the chair to-day. I spoke twice, and one time for Salter's benefit, showing how far he was wrong in his conclusions touching early life.'

As it was now the end of the June quarter, and the accounts had to be gone through, Ramsay came up from Oxford to London, and spent two or three days there, contriving as usual to crowd a good deal both of work and of amusement into the time. Let me cite in illustration the diary of only one day.

'*3rd July.*—Wrought among the accounts, etc., all day, and wrote Willie. At three went to Lady Shelley's breakfast party. There was a most brilliant assemblage. I, however, only knew Lady Shelley, Barlow, and Sir Philip Egerton. I was especially delighted with the children. There were about fifty of them. They looked so lovely, and were so elegantly dressed. A harp and piano were brought out to the green, and the children danced so gracefully. I left at five to join the Red Lion dinner. We entertained the Prince of Canino and several others, Dutch, Russian, and Danish, making no difference in our ordinary fare of beefsteaks, kidneys, toasted cheese, etc. We had two jolly bowls of punch brought in after. Cooke Taylor was in the chair, Forbes vice. Sir Henry De la Beche sat opposite the Prince, on Taylor's left hand. The foreigners of northern nations entered into the fun with heart and soul, and though the nephew of the Emperor Napoleon was wondrously good-natured and seemed highly amused, I question if he perfectly understood the humour of the thing.' By the 8th of the month Ramsay had joined a large Survey party assembled at Cerrig y druidion, from which centre they were working out the structure of the country to the north and east of Bala. On the 9th he makes the following entry :—

'Held a consultation over some trappy specimens in the morning. After that went out with Smyth, Jukes, Selwyn, Gibbs, and the dogs, to look at Jukes's ash-beds. We had a goodish day's work on bits of detail. Jukes should have showed me something larger, but detail seems his forte. He is ever in doubt, even when nearly convinced, about little things, and yet grasps the subject so well notwithstanding, that he produces better work and understands it better than any man on the Survey.'[1] Next day one of the canine companions of the party, 'Jukes's dog Governor, amused himself by slaying a sheep, which cost his master 7s. 6d. In the course of a long walk we got well drenched going, dried again coming back, and were re-wet before we got home.'

De la Beche was at this time in North Wales with his daughter, taking a share in the field-work by tracing some of the boundary-lines of the igneous rocks of Caernarvonshire. He asked Ramsay to join him at Beddgelert, and to accompany him in a kind of prospecting excursion through the country around Snowdon, and thence into Anglesey. This was the first introduction of the younger geologist to that intricate piece of geology, in the ultimate unravelling

---

[1] The reader, as already mentioned, will find an interesting record of the friendship between Ramsay and Jukes in the *Letters* of the latter, likewise glimpses into the life of the Survey men in North Wales.

of which he was to achieve one of the great successes of his career. He probably never spent three happier weeks than these. The beauty of the scenery entranced him, he became more familiar with his worthy chief, and what he always counted a great additional pleasure, he passed the time in cultivated and agreeable female society. On the first day, as they were driving over to Llanberis, which was to be their chief headquarters, he notes: 'To my great surprise and delight, Sir Henry proposed that I should occupy the same quarters with them—have a bedroom, and all mess together. Wasn't I satisfied? The thing was so unusual, no one having ever penetrated before into the sanctum of the family.' A few reminiscences of the tour may be quoted.

'*14th July*.—After breakfast we all started for the top of Snowdon, the girls walking by the road, and Sir Henry and I cutting a parallel section of the Barmouth sandstones, etc., on the neighbouring ridge, and every now and then coming within sight and hailing them. By and by we joined them. It was a glorious day. First, all the country was partially enveloped in white fog, which, clearing off here and there, showed peeps of the country, as if set in a superhuman frame. By and by it all rolled away, and from Cader Idris and Plynlimmon to the Longmynd all was clear and distinct. Confound the Cockney tourists, though, that one meets a-top, and confound the huts and coffee-pots, visitors' books and guides.

'*15th*.—We all started after breakfast for the lake, and got into our landlord's boat, Sir Henry and I pulling, and the ladies laughing and chatting in the stern-sheets. I never enjoyed a day more all my life.

We paddled along and admired the glorious scenery of the lakes, and the Pass of Llanberis, with Snowdon in the clouds, and that old grey tower below. Then, ever and anon, Sir H. and I landed to tap the rocks, chaining the boat with its fair freight to the banks till our return. We pulled to the bottom of the lake, and walked a mile farther, picking the ladies up on our return, as well as a lot of cockles Sir H. had bought. We loitered to gather a hundred or two of white and yellow water-lilies. We then tied a shawl to an oar for a sail and crept up the lake, dragged the boat into the other lake, and so home at half-past six.'

In descending from an excursion to the top of Glydyr fawr he sprained his foot—an accident which, though he made light of it at the time, proved serious enough in its effects to prevent him from further field-work for some weeks. Much of the subsequent occupation of the party was done by driving from point to point, so that the disabled geologist had an opportunity of taking a general survey of the whole region. In this way they visited Caernarvon, struck the southern coast at Pwllheli, crossed to the west side of the peninsula, drove to the promontory of Aberdaron, and thence back by Tremadoc to Llanberis. In these preliminary traverses, favoured by good weather, Ramsay was enabled in some degree to grasp the physical features and broad geological structure of a region into the detailed study of which the Geological Survey was now about to enter. There is an additional interest in these excursions, for one of them included a visit to Anglesey, where Ramsay now saw for the first time the island about which in after years he was to think and write so much, where he was to find his wife, where, wearied with the turmoil of official life, he

was to retire and spend his closing years, and where,
at last, he was to be laid at rest for ever. The party
drove to Beaumaris in an open barouche. The diary
records how they waited half an hour to see 'that
glorious work, the Menai Bridge. Its beauty, simplicity, and grandeur are wonderful. Dined at the
Bulkley Arms, Beaumaris, and after dinner removed to
our new lodgings in Menai Place, which are very nice,
and have a splendid view across the straits.'

This happy time came to an end on the 6th August,
when the De la Beche party left by steamer for Liverpool, and Ramsay, still lame, made his way by carriage
to join Selwyn at Ffestiniog. With the help of a pony,
he was able to accompany Selwyn, Playfair, Jukes, and
Gibbs over a good deal of ground, and discuss with
them some of the problems that had been met with
in the course of the mapping. But as the sprain continued to give a good deal of trouble, he at last went
over to Dolaucothi, and remained six weeks there, to
rest and work at the preparation of his lectures for
University College. Forbes joined him, and the two
friends had long consultations over the general plan
of these lectures. Thus, under date the 25th August,
Ramsay records: 'Arranged with Forbes a plan of
my introductory lecture. By his advice I simplified
and condensed my plan, but I much fear it will be
more than I can well do to make a good job of it, considering the little time I shall have in London to prepare a good philosophical account of *how* folks arrived
at their geological conclusions from the time of Strabo
down to our own date.'

The month of September, and nearly all October,
were spent in taking Forbes over some of the sections
that best showed the characters and relations of the

various members of the Silurian series. From Dolaucothi the two geologists made their way by Llangadoc
and Llandovery to Builth, where they saw the Carneddau, with its unconformability, then by Pen y bont,
Kington, Ludlow, Church Stretton, Bishop's Castle,
and Chirbury to Welshpool. In this tour Ramsay,
still unable to use his foot, was compelled to ride,
while Forbes walked at his side. Being familiar with
the ground, however, he was able to point out all the
salient features of geological structure. ' I explained,'
he says in his diary, ' and Forbes believed in all the
geology.' As Forbes had not had any opportunity of
making himself familiar with the older Palæozoic rocks,
it was of great benefit to him, and of much ultimate
advantage to the Survey, that he should learn his
lesson in such a typical region, and under the guidance
of the best stratigrapher on the staff.

There was not, however, always perfect agreement
between the two travellers. So long as only the facts
of geological structure were concerned, Forbes was
quite content to take them from Ramsay, but when
it came to the interpretation of these facts, and to
theoretical deductions from them, he claimed to use
his own judgment. Notwithstanding the experience
gained in mapping the Cader Idris country, and in
traversing the Arenig chain, Ramsay still retained,
and indeed maintained to the last, his belief in the
conversion of stratified rocks, through the contact
metamorphism induced by intrusive masses of igneous
material, into substances that could not be distinguished from true igneous rocks. He supposed that
the sedimentary strata had been actually melted, and
that from this molten condition a gradation could be
traced, on the one hand, into the ordinary character of

sediments, and, on the other, into undoubted eruptive
material.   After spending some time on Caer Caradoc
and Hope Bowdler he went to Bishop's Castle, and
on the first day of the stay there makes the following
entry : ' 1*st October*.—Went out to take a turn on the
traps and altered rocks at Upper Hublast (*sic*).   Found
that they had been injected into and highly baked the
Wenlock shales, and in one place, as I thought, fairly
melted them, so that part of this must be mapped trap.
Forbes and I had a tough argument on this head, for
I fancied I could trace a gradual change from the
genuine baked Wenlocks into melted beds.   This he
would not allow, so we both became hot, and neither
gave in.   We were doubtless partly both right.'

Edward Forbes was an excellent artist.   He could
with great rapidity catch the likeness of any one whom
he wished to portray, while his poetical temperament,
his vivid imagination, and his keen sense of humour
enabled him to convert his likenesses into idealised
portraits or comical caricatures, as the impulse moved
him.   At Dolaucothi he made pictorial contributions
to the ladies' albums, in which the various members of
the household figured.   His landscape sketches were
likewise often admirable.   His artistic eye enabled
him to seize and delineate accurately the general effect
of a scene, while his geological knowledge helped to
guide him in expressing its dominant features.   Ramsay,
though not so gifted in this respect, was not without a
measure of artistic capacity.   His early drawings of
Arran scenery were remarkably good, and his note-
books contain many characteristic sepia-sketches of
the landscapes through which his official duties led
him to wander.   On this tour with Forbes, not being
able to walk, he seems to have consoled himself by

sitting down to sketch. He makes reference in his journal to these pictorial efforts, but it is generally in some such form as, ' Forbes made some excellent water-colour drawings; I spoiled some paper.'

Among the tastes which these two comrades had in common was a love of antiquities. Ramsay up to the last was always willing to go a long way round for the purpose of visiting a ruined tower or crumbling abbey. He would become enthusiastic as he reconstructed in imagination the design and details of the architecture, and traversed every nook and corner of the ruin, while sometimes the proofs of ruthless destruction would fill him with sadness. Referring to another part of the country, he enters in his journal: ' Revisited all the ruins, got to the top of the square tower, and half broke my heart with the contemplation of such glorious structures utterly destroyed.' On one of the excursions from Church Stretton his antiquarian soul was stirred within him as they traversed the old Roman road, Watling Street, and found it ' now so overgrown that it is a mere grass walk between hedges and briars.'

On the 26th October, after a pleasant tour of five weeks, Forbes went back to London, and Ramsay started for some weeks of hard field-work in Montgomeryshire. A good deal of that country had already been mapped, but there were some parts of it which, from his more recent experience among the volcanic rocks, needed revision before publication of the maps. Accordingly, he devoted himself to the task of re-examining and completing the geological lines, taking long expeditions, getting over a large tract of ground, and definitely fixing some important points in the geological structure of the region.

On the 6th November he writes : ' Out on the hill
south-east of Llandegle Rhos. Could not make a
start without much scenting about and doubling back
and forward. But during this process I lighted on a
glorious sight, proving beyond a doubt all my asser-
tions about the geology of the country. Sir H. and
Smyth ought to have inferred the same when they
mapped these traps. I found Wenlock shale containing rounded pebbles of trap and slate resting
unconformably on Llandeilos on the east side of the
Builth traps. It ravished my soul with joy, and far
more than atoned for the little that was done before.'

' 11*th December.*—A tremendous day's work down
the middle of the traps to the ground above Llanilwidd,
near Builth. Found in Sir H.'s mass of *greenstone* on
the east lots of *fossils !* Ran across the country as far as
Pencerrig, and walked back what is called eleven miles
in two hours and a quarter.' In this traverse he 'put
the finishing touch to the Builth section.'

All the daylight, and sometimes part of the dusk,
in these autumnal days were spent in this active
pedestrianism. But this work represents only a part
of the Director's industry at that time. He kept up
with singular regularity and promptitude a correspondence which, both with his colleagues and with
friends at a distance, was every year growing more
voluminous. He had made some progress in Welsh,
and he used to employ himself in translating Welsh
songs into English rhyme. Nor did he content
himself with mere metrical translation. He had
always been rather fond of turning his thoughts into
verse, and he occasionally penned an ode or sonnet,
or a rhyming epistle. To his good friends at Dolaucothi he often chose a metrical way of expressing him-

self. Thus, on one occasion, with the music of the Spenserian stanza in his ear, he wrote :—

> Or if mayhap, with radiance clear and bright,
>   The morn give token of a goodlie day
> To lure the luckless geologic wight
>   Once more o'er dale and breezy down to stray ;
> Then let him walk and work, while work he may,
>   Forthy eftsoons, though much against the grain,
> He by and by, slow wending on his way,
>   Ah, hapless wretch ! returneth home again
> Bemired above the knees, and drenched with pelting rain.

At another time, after section-running in very bad weather, and getting back 'drenched with pelting rain,' he found consolation in making fun of the discomfort, and with recollections of *Marmion* and the *Lady of the Lake*, penned the following account of it :—

> Dear Emily, two years ago
>   Methinks I promised to send
> Some sunny day a line or so
>   To you, my trusty friend.
>
> Loud howls the wind across the waste
>   Sharp falls the pattering rain,
> And yet I don my togs in haste,
>   And take to the hills again.
> With compass and clinometer,
>   And hammer, stout ally,
> I tramp away from Rhaiadr,
>   Adown the winding Wye.
> What reck my fellows of the rain,
>   William and Thomas hight,
> Who bear the lengthy legs, the chain,
>   And the theodolite ?
> Stout Thomas was a Builth man bold,
>   William was reared to wait
> On John ap John ; his sire doth hold
>   His post beside the gate.
> But when we came to Rhos-saith-maen
>   And set to work, I wist,
> Our merry toil we scarce began
>   When the white and curling mist
> Came downwards like a mighty shroud,
> And wrapt the hills in one vast cloud,

I

So thick and close, you scarce could see
  Within that villain fog,
I vow to good St. Jeremy,
  A chain's length o'er the bog.

But hark ! upon the bleak Drum-ddu
The roaring winds rush fast and free,
And the damp mists that hide the day
Upon their wings they bear away,
And up against the cold blue sky
Stands Cefn-y-gamrhiw sharp and high.
' Now to't, my merry men, like fun,
And make your hay while shines the sun ;
You, Thomas, cut along like wind,
And, William, follow up behind :
Run like old Scratch, my lads,' quoth I,
And off we go right merrily
By stock and stone, nor stayed to breathe
Till o'er the hill and o'er the heath
We reached the vale at set of sun
Where wild Cwm Elan's waters run.

Remote Cwm Elan ! well I ween
I never saw a fairer scene.
Thy sparkling waters winding stray
By meadow green and mountain grey,
  Beneath whose shaggy crest,
In many a wild romantic nook,
By mossy stone and mountain brook,
  Image of quiet rest,
In many a lone and shady spot,
Curls the blue smoke from lowly cot.

Next morning we were boun' to climb
Black Craig-y-foel's cliffs sublime;
  So steep this hill, so tight and tough
To speel, it cost a whole hour long ;
  And ere we reached the summit rough
It cracked my wind and stopped my song.

Then, O kind-hearted Emily,
Most fervently I beg of thee,
Remember in thy nightly prayer
  Thy broken-winded A. C. R.

For some of his old Glasgow friends he chose a
ruder verse, as in the following piece of doggerel :—

I am a geologic tramp
(Beef and greens make very good cheer),
Over the country I rove and ramp
(And a pewter quart is the dish for beer).

I run and I ride, I ride and I drive
(Capon and sausage are good i' the mouth),
Come home to dinner at half-past five
(And a glass of stiff grog will quench the drouth).

I live in an inn by the turnpike road
(Ginger and pepper will tickle the chops),
In rainy weather a queer abode
(When ye brew, i' the brewst put plenty of hops).

You've got a wife and I've got a hammer
(Brose and butter and porridge and ale) ;
Both at a time can kick up a clamour
(And a joke's a joke, though never so stale).

A wife is better in palace or hovel
(O but a blushing maid looks winsome !)
Than a poke in the eye with a dirty shovel
(But a maid looks best when her pocket is tinsome).

I'm unco vexed that I canna gang down
(Up frae the Broomilaw up in a noddy),
But I maun prepare for the U.C. gown
(Breakfasts and dinner, and O ! the toddy).

But besides an increasing, though often amusing and interesting correspondence, there were now looming grimly in front of him the lectures which in a few brief weeks he would have to begin in London. It is not easy after eight or ten hours of active pedestrianism, followed by a good dinner, to sit down calmly to serious literary exertion. Ramsay records now and then with remorse that sleep got the better of him, and he made no progress with those lectures. Nevertheless he set himself resolutely to face the task, and seems to have finished a number of lectures, or at least the detailed notes from which they would be delivered, before he returned to London. For the main part of the course he knew he would be compelled, in a kind of hand-to-mouth way, to work up on one day the lecture that he was to deliver the next.

Not only were the lecture notes to be prepared, but the diagrams to illustrate them all required to be designed and drawn, for there were no appliances of this kind at University College. Ramsay always showed much skill, and even what might be called artistic feeling, in the drawing of geological sections. He now made drafts of what he would need for his course, and sent them up to his colleague, W. H. Baily, at Craig's Court, to be enlarged into proper lecture diagrams. The occasional wet days that interrupted mapping allowed more steady progress to be made with these preparations for the professorship, and once in the full swing of work he would continue until long after midnight, when sleep, which overtook him when uncalled for, would not come when desired, even although he 'read *Count Grammont* for an hour to get rid of the geology on the brain.'

By the 20th December Ramsay was once more back in London. Survey duty kept him busy all day at the Museum, and his lectures still occupied him all evening, and sometimes far into the night. A few of his jottings regarding the preparation of these lectures in town may be given here. 'Made a complete abstract of Steno's *Prodromus* before going to bed.' 'Stuck at Hutton's *Theory of the Earth* and Playfair's *Illustrations* all day (Sunday), and before night read all, and made a complete abstract of the latter.' 'Wrote a bit of lecture, read *The Fortunes of Nigel*, and went to bed at one.' 'Wrote a good bit of

my lecture at night. Hutton every day strikes me
with astonishment. Lyell does not do him half
justice.' He had never had any practice in public
speaking, and was uncertain how far he could trust
to notes, or how much he ought to write fully out.
But he possessed the best qualification for a successful
lecturer : he was full of his subject. To wide reading
in it he could bring the priceless advantage of that
personal acquaintance and vivid perception which
years of practical work in the field could alone have
given him.

A professor's first course of lectures is always
the most arduous. The preliminary gathering and
arranging of notes, and the planning and execution
of diagrams and other illustrations, leave him generally
prepared for, at least, the first few lectures, perhaps
for the larger proportion of the series. But he is
probably seldom able to get all his material in hand
for the completed course before he actually begins
to lecture. Most usually he comes to the end of
his arranged notes when there is still a formidable
part of the term in front of him, and when, therefore,
he has to sit late and rise early to get ready for the
prelection of each day as it comes. Then there is
the feeling of uncertainty which arises in his mind
as to his facility of expression, when, for perhaps the
first time in his life, he finds himself addressing that
exacting audience—an assemblage of lads, many of
them much readier to seek amusement than instruction,
careless yet critical, who have to be attracted and
interested before they can be instructed.

With but little knowledge of students and student-
life, with scarcely any previous practice in public
speaking, and with no experience in teaching, Ramsay

looked forward with some misgiving to the fateful 14th January 1848, when he was to enter upon his new educational duties. He chose as the subject of his Inaugural lecture a sketch of the progress of geological science, selecting a few of the greater names in the bead-roll of geology, and dwelling more particularly upon the labours of the illustrious Hutton.[1] It was a wide theme for a single lecture, but the author succeeded in giving prominence to some of the main historical facts in the evolution of geology, and he reserved for his opening lecture next year a continuation of the story in its progress from the time of Hutton to that of William Smith.

On the opening day of his course the new Professor made the following entry in his diary :—

'*14th January.*—Got up betimes and worked at my lecture till half-past ten. Took a cab to University College, reading as I went the ill-written passages of my lecture. By and by Sharpey and Dr. Grant came in. We then went to the Professor's room, where Graham and Sharpey introduced me into a silk gown, and then Dr. Grant introduced me to the audience, which numbered about a hundred. I was pleased to see so many. Dr. Fitton was there and sundry others, Forbes and Oldham grinning at me from the back rows. I felt a little nervous, but got through very well, as they told me, in an hour and a quarter.'

After a month's experience of lecturing he writes : 'I suspect the listeners are better pleased with my matter than I am, and more than that, I daresay

---

[1] *Passages in the History of Geology*, being an Inaugural Lecture at University College, London, by Andrew C. Ramsay, F.G.S., Professor of Geology, University College, and Director of the Geological Survey of Great Britain. London, 1848.

I learn more than they do.' The course came to an
end on the last day of March, on which date the
following memorandum was made. 'Got the com-
posing steam well up and finished the lecture by eleven.
Got through it unusually well, and had a round of
applause when it was over.'

We may be sure that in this first course of lectures
the young professor touched on many questions about
which he was able to lay fresh views and original
illustrations before his hearers, drawn not from books,
but from long observation of nature. His treatment of
denudation and the results achieved by it would be
specially full and instructive. His account of igneous
rocks and the manner in which volcanic phenomena are
chronicled among the older geological systems would
be such as at the time could be found in no published
book or memoir. His description of the structures of
the older sedimentary masses would be marked by
graphic detail, arising from minute practical study of
the subject in Wales. Those who remember Ramsay's
lectures in later years may well believe that these
earliest prelections would not be wanting in that suggestiveness and foresight which were so characteristic
of his style of treatment. In one of his letters to
J. W. Salter (2nd October 1848) he remarks: 'Last
winter I confidently lectured that these [Welsh] rocks
were Silurian, and also that the Grampian clay-slates,
etc., would turn out to be ditto, more or less altered.'

During the winter of 1847-48 in London, besides
his lectures, there were various incidents that helped to
enliven the daily routine of the Local Director's official
duty. Sir Henry De la Beche had been elected
President of the Geological Society, and as he now
took the chair at the meetings, the fortnightly reunions

had an added interest for the members of the Geological Survey.

In his anniversary address on the 18th February Sir Henry took the opportunity of gracefully acknowledging the cheerful co-operation of the fellows of the Society in the work of the Geological Survey. The Society and the Survey were not rival organisations, but were united in the one paramount object of promoting the cause of geological science. It was a former president of the Society who had been consulted by the Government as to the propriety of definitely establishing a geological survey of the United Kingdom, and had urged the formation of such a department of the public service, while many of the members of the Society had cheerfully assisted and encouraged the efforts of the Director. De la Beche, on the other hand, had been for years Foreign Secretary, and was now elected President of the Society. He was thus enabled to give it the benefit of his wide experience and his excellent business habits. The members of his staff, too, took a share in the affairs of the Society, acting on its council, reading papers before it, entering into the discussions, and contributing material to its *Quarterly Journal.* This feeling of mutual sympathy and co-operation has continued to mark the relations of the Society and the Survey down to the present day. The Society has freely bestowed its offices and its honours upon the members of the Survey, who, on their part, have looked with pride upon their connection with the oldest and most distinguished of the geological societies of the world.

On the present occasion Sir Henry was able to announce the satisfactory progress of the Survey, both as regards its field-work and the issue of its maps. He stated that the maps completing South Wales and extending into North Wales would soon be published, and that considerable progress had been made in the Survey and publication of the maps of Herefordshire, Shropshire, Somerset, Dorset, and Wiltshire. He spoke also of the satisfactory advance of the field-work in North Wales and Derbyshire, and referred to the commencement of the publication of the maps of the Irish branch of the Survey now under his direction. But perhaps the most important announcement made by him was that in which he stated that the collections in the Museum of Practical Geology having so greatly increased, 'the Government is now erecting the con- siderable building, which the members of this Society may have observed extending from Piccadilly to Jermyn Street, where these collections, illustrating both the science and applications of geology, can be made properly accessible to the public.'[1]

The planning and erecting of this new edifice occupied much of De la Beche's time and thoughts for several years, and many were the consultations which he had with the various members of his staff on the subject. His scheme gradually enlarged as he found he could carry the Government authorities with him, until, in 1851, he saw the new Museum completed, and with it the realisation of the bold idea to found a great educational establishment which he had aimed at so many years before.

Besides the fact that his chief was now president, Ramsay had the additional reason for attending the meetings of the Geological Society, that he had been elected a member of the Geological Club. This was

[1] Anniversary address, *Quart. Journ. Geol. Soc.* vol. iv. 1848, p. 81.

the dining fraternity to which, as already described, he had been introduced by Murchison on his first coming to London. Founded in 1824, it consists of Fellows of the Geological Society, limited in number to thirty-six, who are wont to dine on the evenings of the Society's meetings, and to adjourn from the dinner to the meeting. The Club thus serves a double purpose; it brings its members into closer and more social contact with each other than is possible in the Society's rooms, and it secures the nucleus of an audience at the evening meeting afterwards. The Society's apartments were at this time in Somerset House, and the Club met in some restaurant in the near neighbourhood. At first the dinners took place at the 'Crown and Anchor Tavern' until that noted establishment was closed in 1847. The Club then moved to Clunn's Hotel, Covent Garden.

Since the removal of the Society's apartments, in the year 1873, to Burlington House, the old-fashioned dining-houses in the region of the Strand have been forsaken for others nearer the place of meeting—a change still regretted by some who remember affectionately the dingy but cosy dens where they used to dine a generation ago.

A few memoranda regarding the Society and the Club occur in Ramsay's diary of this winter. Thus on the 5th January 1848 he notes : ' Dined at the Geological Club, Clunn's Hotel, Covent Garden, for the first time since becoming a member. Selwyn accompanied Sir Henry, who was in the chair; all the rest were Horner[1] and Prevost, so we were but five. The

---

[1] Leonard Horner, born 1785, died 1864; one of the early fellows of the Geological Society, and twice elected its President. One of his daughters became the wife of Sir Charles Lyell.

dinner was splendid and the wine not bad. At the Geological afterwards we had a paper by Nicol,[1] the new Secretary, which he read in a monotonous, drawling, school-boy voice, like some of the old scholars I remember at the Parish School at Saltcoats twenty-four years ago. The paper was good enough—on the Silurian and part of the Old Red Sandstone of the south of Scotland. Lyell, Salter, Greenough, and I spoke. I rose a little afraid, but got on famously before I had said a dozen words, and, as I was told afterwards, gave great satisfaction to Greenough and some others, who liked the Survey style of treating such subjects. I took good care to clench two things; first, that on analogous subjects some papers would be read by the Survey; and, second, giving Selwyn a bit of laudation to the cheering of his heart.' Next day Sir Henry told him at the Museum that he had been 'much pleased with the Geol. Soc. last evening, but said he was afraid I would speak again and remove the good impression made by my first.'

' 18*th February*.—Anniversary of Geol. Soc. Did not get down from my lecture till after the [Wollaston] medal had been given to, and acknowledged by, Dr. Buckland. Sir Henry's address passed off very well. I sat mostly next Darwin. I was elected a member of council. Anniversary dinner afterwards. Sir Henry did most admirably in the chair, turning off all his speeches excellently. Sedgwick made the best speech of the evening. I was called on to return thanks for the Survey; Playfair for the Museum. I got on well,

---

[1] James Nicol, born 1810, died 1879, was appointed Assistant Secretary of the Geological Society in 1847, Professor of Geology, Queen's College, Cork, in 1849, and Professor of Natural History in the University of Aberdeen in 1853. His best-known papers are on the structure of the North-West Highlands of Scotland, the great value of which is now universally recognised.

all save one short hesitation, caused by my being so intent on my first paragraph that I quite forgot the second.   We broke up about eleven, and in the long-run Smyth, Reeks, Bristow, and I had some supper. Got home at half-past three.

'*22nd March.*—Geological Society night.   Dined at the Club.   Sir H. gone, and Moon in the chair. I sat next Prestwich and Austen, and opposite Forbes and Lord Selkirk—all pleasant men.   The last seems most agreeable and unaffected.

'Good night at the Society.   Buckland made a most witty speech.   It was about crinoids; and he began by saying that the debate seemed to him to have "more of a gastronomic than a palæontological character; for all that had been said bore upon the relation of the plates to the mouth and the mouth to the plates."   Forbes spoke well, and to the purpose; so did Charlesworth and Carpenter.   I was glad of this, for Emerson, the American, was there.

'*5th April.*—Jukes and I read papers to-night at the Geological on N. Wales and S. Wales.   Sir H. was in great alarm beforehand.   Jukes read first. Sedgwick was present, and most agreeable and con-ciliatory.   He made a most complimentary speech after.   Lyell ditto.   Buckland was all in favour, but in attempting to quote Scripture made a great mull of it, and broke down, greatly to the amusement of all, especially the Bishop of Oxford.   I lectured rather much, they told me—the natural effect of a three months' first course of lectures.'

Of the two communications from Survey officers read at this meeting of the Geological Society, one was by Ramsay and Aveline, and was entitled a 'Sketch of the Structure of Parts of North and

South Wales.' It dealt chiefly with the succession of the stratified rocks which had been worked out in the country to the south and south-east of the Dolgelli and Bala district, and pointed out the clear evidence of the great unconformability and overlap of the Upper Silurian formations. The other paper, entitled ' Sketch of the Structure of the Country extending from Cader Idris to Moel Siabod, North Wales,' was by Jukes and Selwyn, and showed the relative positions of the various great stratified groups and their intercalated volcanic masses.

These papers are interesting in the history of British geology, inasmuch as they gave the first published outline of the results up to that time obtained by the Geological Survey in North Wales. As their titles expressed, they were merely sketches; and even in that form they were printed only in abstract. The Director-General, as told in the last extract from Ramsay's diary, was in a state of alarm as to what might happen as a consequence of the reading of these papers. This fear arose from a certain timidity of nature which, with all his energy and determination, characterised De la Beche. So anxious was he for the ultimate success of all his wide scheme for a great national institute of applied science, that though he could show fight when occasion demanded, he shrank from taking himself, or encouraging on the part of his subordinates, any action which seemed likely to stimulate opposition. He did not greatly favour the communication of papers by his staff to scientific societies giving the results arrived at during the operations of the Survey. He contended, and with some show of reason, that these results were obtained by public servants at the cost of the State, and were the property of the country, and not of the individuals who made them. Some of the staff, however, angrily resented any restraint of their liberty in this respect; and there seemed at one time the possibility of a serious rupture on the subject. In the angry correspondence which took place between one of the malcontents and Sir Henry, it was asserted that by the course which had been followed Ramsay had been prevented from taking the position as a geologist to which the amount and quality of his work entitled him. This was a charge which gave special pain to Ramsay himself when he heard of it. 'This is too bad,' he wrote in his diary; 'for though by reading more papers I might have stood higher at the Geological Society, yet, all in all, Sir H. has been my best friend in every way, private and public.'

The mutiny had been in progress during the autumn, but as soon as Ramsay got back to London his tact and sound common sense succeeded in not only keeping the peace, but in effecting an arrangement which, while it preserved the due discipline of the service, provided for the officers of the Survey the possibility of making known their observations in anticipation of the subsequent publication of an official account of them. He wrote to Aveline: 'I have achieved a great point, and got permission for the Survey to read papers at the Geological Society, on the approval of the Director-General, when submitted to him by me. I wish, therefore, you would think over some of the particulars of your present trappy country, as I wish much, if you have no objections, to associate you with myself in a joint paper on the Stretton, Bishop's Castle, Kington, and Builth land. It must not deal with details, but be *general*, and yet precise in its conclusions. We must not give sections, but diagrams, such as other folk call sections, showing the general run of things; also only a sketch of the map. The reason is, that as yet the country is unpublished, and matters handed in to the Geological Society belong to it. It is a great point, however, to have gained this, for the Survey is not half enough before the public.'

This was by no means the only occasion on which the Local Director was able to avert such threatened ruptures between the 'officialism' of the Director-General and the 'licence' of the geologists. De la Beche was by instinct an official, and he had lived so long in intimate contact with ministers and departments that his natural bent of mind was intensified. If there were two ways of getting a thing done, he chose the more official and roundabout rather than the more simple and direct. Probably he was generally right in his choice, but to those who looked on from outside, and were not cognisant of all the facts, he seemed often to be raising needless difficulties and guarding against objections that were never likely to be made. Always courteous and pleasant in manner, he seemed unwilling to give a blunt negative to a request, and thus sometimes, unwittingly, encouraged hopes that he did not mean to fulfil.

Ramsay seems to have formed a tolerably just estimate of the character of his chief, whose weak points he recognised, while he thoroughly appreciated his excellences. At the time of the outbreak above referred to he wrote in his diary that '—— has used harsh and even unjust terms to Sir Henry, and ——, too, is not fair to him. Sir Henry's devotion to the Museum and Survey sometimes blinds him to other matters. People must make allowance. With all his little failings, I wish I knew more men I love as much.'

There can hardly be any doubt that the very ' officialism,' which seemed to some of De la Beche's critics a defect, powerfully contributed to his success in gaining from the Government of the country support to his scheme for the national endowment of applied geology. He knew how to measure and influence the official mind. He began by trifling requests, and gradually educated the various departments to adopt his views and give their assistance to carry them out, until it became as much a point of honour and credit with his official superiors as it was a heartfelt desire of his own that his successive demands should be favourably considered at the Treasury. Moreover, he cultivated personal relations with the ministers of the day. He was on specially friendly terms with Sir Robert Peel, whom he led to take interest in the erection of the new Museum and in the progress of the Survey. He even sounded him as to his acceptance of the Presidency of the Royal Society. On the 9th February of this year De la Beche told Ramsay that he had been on this errand, and that ' Sir Robert refused on the ground that it ought to be a scientific man. He (Sir R.) highly approved of the plan of holding the soirées at the Society's rooms ; " and then," said he, " if a poor man, as it might, and often ought to be, held the Presidency, we could go and pay our respects to him." '

But the Director-General and his staff at Craig's Court had other duties to discharge this winter than had ever before fallen to their share. The proposal of the Chartists to assemble 200,000 men on Kennington Common and march to Westminster on the 10th April led to the taking of ample precautions for the security of public buildings in London. Though the establishment at Craig's Court might have been supposed to lie almost hidden away from the ken of any rioters, its officials prepared themselves most manfully to resist the invasion of their premises. These preparations, and the eventful day, are thus chronicled in Ramsay's diary :—

'*8th April.*—Got sworn-in to-day a special constable; got a baton at Scotland Yard. Forbes refused; his usual policy. He says there is no cause for alarm, and yet commends people for taking precautions! Yet *he* takes none. Sir H. also *organises*, yet does not swear - in himself. Playfair, Hunt, Baily, Reeks, J. A. Phillips,[1] etc. etc., plucky.

'*9th.*—Wilson is also sworn-in, and quite ready to do the needful. Quiet enough to-night; doubtful to-morrow.

'*10th.*—Grand row expected to - day. Forbes called, and we went down to the Museum before ten; met Playfair. Sir Henry at the Museum very active and mysterious, passing through holes into the back stables of the Scotland Yard Police Office, and bringing out armfuls of cutlasses. Streets full of special constables. Chartists afraid, and cowed; all passing off quietly. No procession took place. However, we had a jolly dinner in Sir Henry's room for fourteen, and cigars and coffee in the laboratory afterwards.

---

[1] John Arthur Phillips, born 1822, died 1887; received his training at the École des Mines, Paris. At the time referred to in the text he was assistant to Professor Playfair in an investigation ordered by the Admiralty into the steam coals best adapted for the Navy. He afterwards became a consulting engineer in mining and metallurgical matters, and travelled much abroad professionally.

This was the hardest duty we had to perform. On public grounds, our men were well pleased that things went off quietly; but as private individuals, many seemed rather disappointed that there was no scrimmage, especially Bone and J. A. Phillips, who were very bloodily inclined. Salter was evidently in a funk, and kept up his spirits all day by whistling psalm tunes.'

One of the pleasantest interludes of Ramsay's life this winter in London was a visit paid by him to Darwin's hospitable home in Kent, when Lyell and his wife, Owen and Forbes were likewise guests. It was a brief sojourn from Saturday to Monday, of which he records :—

'13th *February, Sunday.*—Rose betimes, had a walk in the gardens, and came in to breakfast. Set to work after, and read and thought over Hopkins's views as shown in *Jameson's Journal,* and when found made a note. After lunch Forbes, Owen, Lyell, and I had a walk in Sir John Lubbock's park, and saw a number of things pleasant to look upon, in spite of a tendency to drizzling. Nice cosy chat, too, before and after dinner. Darwin is an enviable man—a pleasant place, a nice wife, a nice family, station neither too high nor too low, a good moderate fortune, and the command of his own time. After tea Mrs. Darwin and one of her sisters played some of Mendelssohn's duets, etc. etc., all very charming. I never enjoyed myself more. Forbes came to my room before going to bed, and gave me a sketch of his coming lecture on generic centres. Lyell is a much more amusing man than I gave him credit for. Mrs. Lyell is a charming person—pretty, lively, and full of faith in, and admiration of, her husband.

' Mr. and Mrs. Lyell told some capital stories about

America, but on the whole all tending to the honour
of America. He is quite enthusiastic about it, especially in all that relates to the liberal spirit of the New
Englanders. Boston seems in all the world his
favourite city. The worst party in America is the
party that emigrates from Great Britain.'

CHAPTER V

THE SURVEY OF THE SNOWDON REGION

WE have now to enter upon the records of three of the most active years of Sir Andrew Ramsay's life, during which he achieved his chief geological triumph —the unravelling of the complicated history of the ancient volcanic region of which Snowdon forms the centre. Though the details of one working season were closely similar to those of another, the story has so much interest in the progress of the geological investigation of the British Isles, that even at the risk of a certain amount of repetition it will be most appropriate to keep the doings of each year distinct.

By the middle of April 1848 the work of the previous winter was at last happily at an end. It had been an exceedingly onerous time for Ramsay, and he confessed now and then that he had nearly reached the limits of his powers of endurance. Before the season closed he was fain sometimes to shirk an evening reception or discourse and take refuge in the reading of Boswell's *Life of Johnson*, or some other favourite. It was, therefore, with no little alacrity that he packed his portmanteau and started for the field again on the 18th April. From that time till the middle of December, with the exception of a little break to attend the British Association meeting, and a few weeks spent in Scotland, he remained in Wales, working out the geological structure of Caernarvonshire.

Beginning his campaign with a tour of inspection along the south coast, he made his first visit to the Isle of Wight, and spent some pleasant days with Forbes and Bristow rambling along the base of the Dorsetshire cliffs. They had among the incidents of this excursion an experience of one of the difficulties in the life of a geological surveyor in the more bucolic parts of England. They were crossing a farm when the farmer rushed on them with angry execrations and violent flourishings of a large spade, with which he threatened to make an end of them if they did not instantly move off his land. In vain they endeavoured to expostulate and to explain their object. The infuriated tenant only became the more defiant. Next day he had not cooled down, but now swung round a still more lethal weapon, would listen to no remonstrance, and had at last to be brought to his senses by a summons before a magistrate.[1]

After brief visits to Aveline and Jukes in the Breidden district, and tracing some new lines there himself, he passed on to Selwyn at Port Madoc and Dolgelli. Room may be found here for his memorandum of one day in this visit of inspection.

'*8th June.*—Up again to the hills south-west of Craig y Cae. Got in some faults and a lot of strange dykes and spots of squirted traps. Selwyn and I separated and took different ground, and often met again to compare and compile. A lovely day, and the effects

---

[1] By the Act 8 and 9 Victoria cap. lxiii. power was given to enter lands for the purposes of the Geological Survey, and to prosecute any one interfering with the work.

over the valley of Dolgellaw and the towering range of Cader Idris most strange and glorious. At last all the lower clouds (which long hung like a half-fallen curtain in the foreground, behind which the sun gloriously illumined the distant glens and precipices) cleared away, and all along the ridge of Cader and the giant slopes of Aran Mowddy the shadows of scattered clouds flitted by like the images of huge flying dragons. I like this plan of separation and meeting. It is pleasant to get alone among the shattered rocks, where one can soliloquise, sing, and shout at will without any man to think you a fool. Home to dinner at six.'

On the 14th of the same month he took up his quarters at Llanberis, for the purpose of himself attacking Snowdon and the surrounding region. The year before, during the preliminary traverses with the Director-General, he had been able to take a general or bird's-eye view of this picturesque district, and had seen enough of its geology to recognise the extraordinary interest as well as the extreme complexity of its problems. He had determined to devote himself heart and soul to their solution, and now at the earliest opportunity, in full vigour of body and mind, he had come back to carry his resolve into effect. His life and work at Llanberis may be best pictured in a few extracts from his diary.

'21*st June*.—Out north to Marchllyn Mawr. Descended to the lake. While minutely examining this section, and hammering along, out jumped a trilobite and a lingula, some 600 or 700 feet down in these "Cambrians," as we called them. So here at a blow vanishes the idea, which we all believed, that the rocks are unfossiliferous beneath the trappy series. Therefore Barmouth, Longmynd, and Llanberis purple lower ground, if one, still do not present the beginning of life, unless a lingula and trilobite were first called into existence where now reposes Marchllyn Mawr.

' 30*th.*—Out after breakfast to touch up part of the sandstones and make out part of the Snowdon section on the ridge above the Pass of Llanberis. What with the interminglings of ash and slate, I see it will be a matter of extreme difficulty, especially as the rocks are much rolled.

' 1*st July.*—Stormy and cold. Up the Pass of Llanberis. Set to work to trace the steep ridge of Llechog. Up and down twice, and half up and down several times. Steep work, consequently not much to show for it. I climbed up and down places that from the road seemed impracticable.'

These labours were for a brief interval suspended while Ramsay went to Swansea to attend the meeting of the British Association. Under the hospitable roof of Mr. Dillwyn (whose son had married one of De la Beche's daughters), and with Sir Henry himself as a guest in the house, he spent a memorably pleasant week. He acted again as one of the secretaries of Section C, and read a paper ' On some Points connected with the Physical Geology of the Silurian District between Builth and Pen y bont, Radnorshire.'

' 11*th October.*—Splendid morning. Started at half-past nine for the hills at the top of the Pass, and sent Gibbs to search the ridge of Snowdon. Sir H. and Forbes [who had recently joined him at Llanberis] followed about half-past ten for the top. While at work on the side of Crib goch I heard Selwyn's well-known shrill shout, and soon discovered him on the top of a crag on Crib goch. So we joined and compared notes, and soon put matters straight at Glas lyn. We then passed on to the top, often standing to discuss, and just as we got to the bottom of the peak descried our party coming down. We stayed nearly an hour up, and then followed. Forbes was making a bad sketch where the path turns down to Pen y gwryd; Sir H. and Gibbs fossilising.

' 20*th*.—Gibbs and I started at half-past nine up Snowdon. Went down to the copper mine at Llyn-du-r'Arddu. We climbed up the face of the cliff there, just by the great fault—a fearful place. It was frozen over in many places with ice and snow. It took us a whole hour to climb it, and we were frequently obliged to stop when in a secure position to beat our hands to warm them. We had often to cut steps in the rock and ice. Gibbs never for a moment lost his coolness, but I got a little nervous for two or three minutes. Once up half-way it was impossible to return; we were obliged to go up. Had a foot or hand given way one or both of us would have been smashed. Parted on the other side of the ridge. I walked across Snowdon to Beddgelert. The top was covered with snow; fine view. Got to Beddgelert by six, just before Selwyn's dinner.

' 2*nd November*.—Out on the ridge on the north-east end of Crib goch. Sometimes misty, but on the whole a good day. Finished all that side as far up as the upper end of the Pass, and to the brook that runs from Llyn Llydaw. Excellent day's work, especially as it fairly finished all that side of the Pass.

' 15*th*.—Up the Pass and up Glyder by the new path I discovered yesterday opposite Pont y gromlech. This mountain begins to be as familiar to me as Charing Cross, and shows evident symptoms of at length beginning to be licked into geological shape.

RICHARD GIBBS

Had a grand find of large *Orthidæ* to-day in the ashy sandstones above the nodular trap.   Gibbs and I climbed to the summit of that huge tower-like precipice, from which the masses of volcanic breccia have fallen, misnamed a cromlech.   It is a fearful cliff to look down, but wide and quite secure at the summit.'

To geologists, and especially to those who are familiar with Sir Andrew Ramsay's name as a writer on glacial phenomena, and who remember his early descriptions of the ice-work in the Pass of Llanberis, it may be of interest to know that he seems to have been at work for some months in that district before his attention was arrested by its glaciation.   We have seen how he curtly dismissed Buckland's views when these were criticised adversely at the Geological Society.   While he makes many notes about other geological matters observed by him on ground which he was examining for the first time, or mapping in detail, he never alludes to the superficial phenomena which a few years later so fascinated him.   The first reference to the subject in his diary occurs under date 3rd August 1848, on the occasion of a visit of Robert Chambers[1] to him at Llanberis.   It runs as follows : ' Selwyn, Reeks, and Smyth up Snowdon ; Chambers and I out on glacial excursion up the Pass, etc.   Very instructive work.'   Next day he remarks that the party, including Chambers, ' started for Llyn Idwal,

[1] Robert Chambers, born 1809, died 1871, best known for his contributions to general literature, took much interest in science, especially in geology.   He is now known to have been the author of the famous *Vestiges of the Natural History of Creation.*   He especially studied raised beaches (see his volume on *Ancient Sea-margins*) and the traces of ancient glaciers, and in pursuit of his researches in these subjects travelled not only over most of Britain, but into Switzerland, Scandinavia, Faroe, and Iceland.   He wrote numerous papers giving the results of his observations, many of which appeared first in *Chambers's Journal,* and were sometimes separately reprinted, as in the case of his *Tracings of the North of Europe.*

and walked across the hills to Llanberis. Splendid examples of glacial action.' Chambers had come purposely to see the evidence of glaciers in the Welsh valleys, and to compare it with what he was now familiar with in Scotland. It looks as if this visit of his had really for the first time turned his companion's eyes from the rocks themselves to the study of the manner in which they have been worn and striated by ice. Ramsay seems to have been still much in the state of mind so well described by himself a few years later. 'We recollect well the unbelief and ridicule that greeted the announcements of Agassiz and Buckland in 1840-41, that glaciers once occupied the greater valleys of the Highlands of Scotland and of Wales, and how sceptics and shallow wits, whose geology perhaps rarely extended beyond the precincts of turnpike roads, attributed the grooving and striation of the rocks to cart-wheels and hob-nailed boots, and the ice-polished rock surfaces to the sliding of the caudal corduroys of Welshmen on the rocks, to slickensides and sea-waves, and to every cause, indeed, but the true one.'[1]

By the 15th November, however, he had been led to recognise everywhere the peculiar smoothing and polishing produced by moving ice; for on that date, with regard to the summit of the tower-like precipice referred to in the citation above, he remarks that this summit 'is, as usual, well grooved with glacial undulations.' Yet it is noteworthy that these are the only allusions to glaciation in the jottings of his first year's work in North Wales. He had evidently not yet realised the nature and force of the proofs of

[1] Review by A. C. R. of fifth edition of Lyell's *Elementary Manual of Geology* in the *Edin. New Phil. Journ.* April 1856, p. 317 (see *postea*, p. 238).

former glaciers in this country. He had never been abroad. The revelation which the first sight of a living glacier flashes upon the mind of a geologist was still to come to him. And thus we find him passing day after day up and down the Pass of Llanberis, heedless of the ice-worn knolls and perched boulders which he was soon so enthusiastically to visit and revisit, and so lovingly to sketch and map and describe.

It has been the custom for foreign governments from time to time to send delegates over to this country for the purpose of personally seeing how the work of the Geological Survey is carried on, with a view to the initiation or improvement of geological surveys in their own countries, or for other purposes where a knowledge of detailed geological mapping may be desirable. During Ramsay's long stay this year at Llanberis he had two such foreign visits. In June A. Sismonda, the well-known Tuscan geologist, accompanied by a young French friend, was awaiting him in his room one evening on his return, drenched and weary, from a long tramp on the hills, and they subsequently accompanied him to his work in the field. 'Sismonda not being much of a climber,' Ramsay writes, 'preferred the road to the rocky sides of the hills. He is still of the Élie de Beaumont school, believes in prodigious terrestrial actions down to the end of late Tertiary time, working with a force of which we have now no experience — earthquakes shaking, traps heaving, and currents sweeping. At night I got the Frenchman and him into a hot political argument, the Frenchman being republican, the other monarchical. Their animated countenances and rapid gestures were most unlike anything one sees in an English debate.'

In August the advent of two bearded Austrians, with large slouched hats, made some sensation among the peasants of Llanberis. One of these visitors was the distinguished Franz Ritter von Hauer, so long Director of the Geological Survey of Austria, and now head of the great Museum of Vienna; the other was Dr. Moritz Hörnes, a well-known Austrian geologist and palæontologist. They accompanied Ramsay in some of his tramps over Snowdon, and received much Survey information from him for a report they were making to their Government. The diary records the ravenous appetites of the party at the evening meals after long days in the keen mountain air, and speaks of 'ogres devouring fish and legs of mutton.'

Not the least pleasant episodes in the Llanberis life were the occasional visits of members of the Survey staff. Selwyn, who was stationed at Beddgelert, would sometimes work over the hills and spend the evening and night with Ramsay, who in turn occasionally crossed the watershed, and landed in time for dinner at Beddgelert. Edward Forbes, who had recently married, brought his bride to Llanberis, and Ramsay took a room in their cottage while they remained there. But no colleague was so welcome as his worthy chief. On an October evening a car arrived at Llanberis with luggage, but no traveller. Ramsay,˙ however, recognised the old portmanteau, and, sure enough, immediately after up came Sir Henry 'shouting and making as much noise as possible.' They had long consultations together on Survey plans and prospects, and one Sunday the Director-General became specially communicative to his younger associate. The conversation is thus reported: 'A walk in the light rain with Sir H., more than usually agreeable. He was very kind and confidential, speaking in the strongest manner about his wish that I should succeed him, and recommending me to write some good memoir speedily for our work, to strengthen my case. "It is not Phillips," he said, "nor any other man on the Survey you have to fear, but such as Murchison and Lyell, who would make an effort. Lyell has so often of late asked me how I did this and that, that I begin to be suspicious." He further said he would try to get an increase of pay for me, and that independently of Oldham, on the ground of my larger charge. I said I would fain see the others with larger pay. He replied, "You must have it first."'

How cordial the relations were between the chief and his lieutenant may be gathered from two notes of De la Beche of this period :—

LONDON, 18*th November* 1848.

MY DEAR RAMSAY—It is refreshing and a comfort to get letters from your honest self, instead of some that I do receive, and from those whom I have laboured to benefit. I even got one three or four days since, containing a passage which looked marvellous like a charge of impeding your fair fame. At least, I cannot make anything else of it. But, mind you, this is strictly between ourselves.

You give a capital account of yourself and your rocky parliaments, making me long to be climbing the hills instead of wending amid sooty streets. However, I believe I am usefully here for the good cause; for the new building is getting on famously, and, among other things, the lecture-room has turned out famously as to light, sound, and accommodation-space.—Ever sincerely,

H. T. DE LA BECHE.

LONDON, 7*th December* 1848.

MY DEAR RAMSAY—Yours rejoiceth the cockles of my heart. Those great 'DONES' of yours were right welcome, as is also the intelligence that you will be shortly up here. I have much to consult my geological son about—fossil proceedings, etc. etc. . . .— Ever sincerely,   H. T. DE LA BECHE.

As reminiscences of the winter season of 1848-49 in London, a few jottings from Ramsay's diary may be inserted here. Besides the completion of their official map-work and memoir-writing, the geologists of the Survey were wont to signalise their assembling in London by a dinner, where they wore their official buttons and sang songs which were written by them for the occasion. Of the earliest of these annual gatherings no continuous record has been preserved, but from the year 1850 onwards the original songs have been entered in 'Ye Recorde Boke off ye Royale Hammereres, off whyche Anciente Ordere Tooballcane and Thorr were erlie Knyghtes.' The subjects chosen for these metrical effusions generally bore reference to some of the work that had been in progress during the previous year, or to some incidents in the life of some of the staff. For a number of years Ramsay never failed to bring his contribution to the hilarity of the after-dinner minstrelsy, sometimes producing as many as four original songs, and singing them with great vigour. Some of these compositions will find a place in later pages. The chronicle does not show that De la Beche ever ventured into rhyme, though he figures prominently in many of the songs. But his successor, Murchison, used to write, and, to the best of his ability, sing his song at the annual dinner; while Forbes, Smyth, Jukes, Salter, Baily, and many of the later members of the staff were frequent rhymesters.

The dinner this year (1849) was held in Covent Garden, and Ramsay records of it: 'We sat down some twenty, Sir H. in the chair, Oldham vice. A right jolly dinner; some capital songs, all original; Salter's and Smyth's best.'

The meetings of the Geological Society are briefly noticed in the diary. Thus under date the 3rd January we get an amusing glimpse of the Council: 'Geological Council to-day. Tough fighting about the Museum Committees. Greenough at five began to speak, and said he could not speak for less than an hour. Dismay reigned. However, he was stopped, and the debate adjourned. Club dinner after; small but pleasant party. I sat between Sir Charles Lyell and Forbes. So-so night at the Society after. I spoke a few words on the Ridgeway cutting. Sir Roderick Murchison was there—the first time I have seen him for nearly two years. He has given up the wig on the Continent, and looks much better in consequence.'

Sir Henry's tenure of office as President of the Society would terminate at the anniversary in February, and Lyell had been nominated as his successor. The new President takes the chair at the annual dinner which is held on the evening of the anniversary, and it is his part to invite such official or other guests as he may wish to be present. Lyell had now this arduous and troublesome duty to discharge. Ramsay writes under date the 10th February: 'Lyell with us a long time, anxious and waiting. He is beating up prodigiously for big-wigs to attend the Geological dinner, and will be miserable unless Sir Robert Peel be there. Sir R. ran over the new Museum this morning with Sir H. and Dr. Buckland. He was (says Sir H.) "charmed." He said the building of it was an act performed in his administration on which he could always look back with pleasure.'

The anniversary of the Geological Society took place on the 16th February, when De la Beche gave

his second and concluding annual address before
vacating the Presidency. In this discourse he
announced his expectation that the complicated district of North Wales would be completely surveyed
during that year. In this hope he made rather too
little allowance for the excessive and difficult detail
which the area contained, for it was not found possible
to finish the region until the summer of next year.
He referred to the publication of the maps of Cardiganshire and Montgomeryshire, and to the fact that
those of other parts of North Wales were in the hands
of the engraver. Dorsetshire and Derbyshire were
nearly completed, and the mapping of the Tertiary
deposits had advanced into Hampshire.

Ramsay's account of this anniversary meeting was
as follows : ' Sir H.'s speechifying day—the Geological Anniversary. Prestwich was awarded the
Wollaston medal. In rising to present it, Sir H.
upset two large oil-lamps that stood on the table
before him and made a prodigious smash. All the
house laughed, and poor P. was a trifle discomposed.
He has a glorious head. Sir H.'s speech was said to
be excellent. I was obliged to run off to lecture.
Went down from College to the dinner at the
Thatched House Tavern. I sat betwixt Playfair and
Captain James. Reeks, Bristow, Smyth, M'Coy,
Tylor, Austen, Forbes, and I were all in a lump.
Lyell made a poor speaker in the chair. Sedgwick
made a magnificent speech, the Archbishop a
goodish one, Van der Weyer a good one, Sir H. a
good one, Buckland a fair, Sir Robert Peel a splendid
one, Murchison an indifferent one, from trying too
much.'

Ramsay continued frequently to attend the Royal

Institution Friday evening discourses. He thus chronicles the evening of the 9th February : 'Went to the Royal Institution to hear Owen on Limbs. I stood on the steps. The lecture seemed to be admirable. Much of it I highly admired, and much of it I did not understand. The theatre was quite full. I saw many I knew : Dr. Fitton looking good-humoured, Sir Roderick looking anxious to keep awake, Dr. Mantell looking eager, Dr. Macdonald looking jolly and anxious for a hole in Owen's coat, Sir Henry looking attentive and queer when Owen came to the orthodox peroration, Sir Charles and Lady Lyell looking knightly, Lady S—— looking vulgar, Nicol looking Scotch, with a doubt in his eye, and Mrs. F—— looking at her dress.'

The Red Lions kept up their London dinners, which were sometimes specially mirthful. Thus on the 19th April Ramsay writes : 'Walked over to Anderton's with Reeks to dine with the Red Lions. Capital party, Lankester in the chair. I sat between him and Sheean, a barrister, and the great original of the Mulligan of Ballymulligan. He seems a capital fellow, though, and sang some excellent songs. Turnberry sang well, and put the whole table in a roar. I scarcely recollect a better evening. Owen was capital, and made a most humorous speech, contrasting the pleasure of sitting in this snowy night, so cosy and merry round the table, with the horrors of the Royal Society then sitting, where the members, on cold benches, in a room with newly-lighted smoke-belching fires, sat listening to a dull paper, with the prospect of one still duller before them. Percy enjoyed himself in his usual hearty style.

Of the dinner-parties and receptions, room can be

L

found here for the mention of two only. ' 18*th February*. — Sir Roderick Murchison's dinner at seven. When I walked into the drawing-room Lady Murchison came running up to me with both her hands out, and made me sit down beside her. . . . Sedgwick was there, Pentland, and Lockhart, Sir Walter's son-in-law. I was delighted to meet him. We had a capital evening. Lockhart was most amusing and interesting. He told a strange story of Lord Brougham, who, it appears, never goes home from any party without first going and taking tea with Lola Montes! I wish I could recollect half the things he said. He is a thorough man of the world and of society, and most gentlemanly, though a trifle abrupt in manner. I did not altogether like the way he spoke of my old friend Dr. Chalmers and his posthumous works.

' 23*rd March*.—Went to Barlow's. A crowd there; among others Dilke and his wife, Baden Powell and his wife, Lady Shelley, Miss Grant, Captain and Mrs. Smyth, Warington and Miss Smyth. Louis Blanc! Some ladies made a demi-lion of him. I was ashamed of them, and wondered Barlow could ask such a man to his house. I would be ashamed to have so foolish and mischievous a fellow in mine. He is a little pragmatical individual, insignificant in person, and insignificant in any appearance of an enlarged intellect. *Petitesse* is the word that expresses him in all things.'

In the prospect of soon taking the field again, he wrote to Aveline from London on 27th March :—

My dear Talbot—My lectures will be over this week. I shall examine the class on Tuesday, and as soon after as possible, that is to say, when I have got rid of Gibbs and the fossils, I shall fly to the country. It will probably take me all that week after Tuesday to finish with Gibbs. Then I join Jukes for a few days. Thereafter I shall go to the Shrewsbury country, principally to look at the Silurians and traps that Smyth traced in, before publishing the map. A few days should do that. I then purpose taking you by storm on my way to Caernarvonshire, so that I may see what sort of strange ground you are on, and also that we may hold a grand geological palaver. I fancy it will be well-nigh the end of April ere I can reach you. Where do you think you may have progressed to by that time?

But it was the usual fate of such prospective plans of work that they could not be carried out within the specified time. It was the 20th April before Ramsay could leave London. He first joined Jukes, who had been at work in the Staffordshire coal-field, and who was now about to run some horizontal sections in the Dudley district. These two friends were becoming every year more closely knit together in intimate friendship. Ramsay, for instance, writes: 'Jukes rises daily in my esteem; he is a noble fellow.' It was while this Midland work was in progress that the official intimation reached Ramsay of his election into the Royal Society. As far back as the 21st April he had heard from his kind-hearted chief that he was one of the fifteen candidates selected by the Council. He might well regard himself as fortunate in reaching this honour after not more than eight years spent in the active prosecution of scientific work.

On the completion of the section-running with

Jukes he once more made some critical traverses across the Wrekin country, and it was the 20th June before he found himself back at Llanberis to resume the survey of Caernarvonshire. Some extracts from his diary and letters will show the nature and progress of his occupation during the campaign of 1849.

'*27th June.*—A jolly day on Glyder; clear but cold. Got a clearer notion of things to-day than I had in weeks of work towards the close of last year among the fogs.

'*29th.*—Y Glyder fawr; glorious day, but extremely warm. Scarce seem to have made any impression on it yet, it is so tough and difficult to climb.

'*30th.*—Across the hills by Mynydd Perfedd, nearly to the Ogwen, and from thence making out the section up to Twll-du—a most rough and craggy walk. A glorious day, which I perfectly enjoyed. Lunched on the banks of Llyn Idwal. Then scrambled up to Twll-du, as far up the gap as I could go—full of rare rock-plants. Thence I scrambled up the cliff, and got home by half-past six. Found twelve or thirteen letters.

'*6th July.*—Took horse and rode to Caernarvon [to have the accounts sworn-to before a magistrate], and got them off to Reeks. As I rode home I found them busy on this side of Caernarvon sinking for coal. I hallooed to a man to hold my horse a moment while I ran into the field and talked with the sinkers, etc. They have gone down seventy yards or so, the first seven yards in drift. They asked my opinion. I told them to let me know when they came to the coal, and I would come down and eat it.

'11*th*.—Over the hills tracing the Bwlch-y-gywion trap, and so back by the felspar stuff up some hideous banks. It was exceedingly fatiguing, but I got a good day's work done.'

LLANBERIS, 12*th July* 1849.

MY DEAR TALBOT—At length since Monday last we have had fine weather here, and I have worked so hard that I am quite fatigued to-day, and stay at home to despatch some maps, and knock off the arrears of correspondence. I think the ground I am at present at work on is really the most fatiguing I have yet experienced in Wales. It is not merely walking up and along steep places, but actual climbing, hands and feet, and on hills so high that it often takes two or three hours to get to the district in the first instance. I fancied ere I came I should be done ere this, but I haven't more than a half or two-thirds finished yet.

'16*th*.—Started at half-past ten, and by dint of sharp walking was at Twll-du by twelve. Down to Llyn Idwal, and traced all the lines round and through the lake and down to the lower margin of Llyn Ogwen, and then up by the Pass-y-benglog and the west ridge of Cwm Bochlwyd, tracing a line to the top of Y Glyder fawr. It was dreadfully tough work, and it was past six by the time I got to the top of Glyder, so that though I would fain have carried on my line, I was somewhat tired both in the legs and of the subject, and therefore deemed it wiser to leave its prosecution for a fresh day. Overtook a nice-looking young fellow in the Pass with a knapsack on his back, and entering into conversation, we walked down together. It lightened the way a bit. Dined at nine.

'26*th*.—Immediately after breakfast started on a

long tramp round by Capel Curig way, tracing the outside boundary of the Glyder fawr trap, and intending to come home over Trefan. But it was too far, and, besides, the work would lead in another direction. So I came back down that rough hillside above the lake and Pen-y-gwryd. It is a terribly stony place. I got into the Pass about six, and was shortly after right well pleased to spy a large two-horse return car coming down the road. Jumped therein. Just about Pont-y-gromlech heard a shouting, and looking up the side of Glyder, saw all my fellow-lodgers and Dent rushing down the hill. They all got in or on the affair, two hanging on behind like footmen. So with mickle laughter we drove home to dinner.

'*6th August.*—As I could not sleep quiet in my grave had I not been up Snowdon, to see that bit on the Beddgelert side of Cwm-y-Clogwyn that bothered Selwyn and me so much, I revisited it to-day, and came back over the top. No one was there but myself.

'*10th.*—Started from Llanberis at nine. Met a Capel Curig car, and changed into it at the top of the Pass, and was at work by half-past eleven or twelve on this side of Y Glyder fach. The mist persecuted me dreadfully. It came rolling down as soon as I got up a considerable height, and then, when I began to descend a little, would partially clear up; but rushing down again, I was forced to try the section on the low ground, and then having made out a certain amount of that, I traced a line up the hill. No sooner had the mist got me well up than, shifting his quarters, he rushed down the valley, obscured Y Trefan, thicker and thicker, boiling and seething, and if I but looked at a bit of ground, down he came upon it and enveloped my head in the mist. At last I was fain to leave about seven. When once I was well down in the valley the white clouds all cleared away from the hills, as far as I could see, though when once or twice I looked back with a speculating eye, I could just see the hill-tops suddenly get partially obscured, as if old Kuhleborn were saying, " You needn't come here, young man, or I'll be down upon you in no time." Got home to the inn about half-past eight, and had a " rough tea."

' 11*th*.—Started after breakfast and began to trace lines from Y Trefan up to Y Glyder fach. Just as I got to the top of the ridge, a gale of wind came on, accompanied by a deluge of rain and a thick mist. I couldn't see thirty yards. A compass was nearly useless, for the ground was so rough that I could not walk in a given direction ten yards, and the place was cliffy on sundry sides. By and by, calculating how the wind blew, I turned my face to it and began carefully to descend, and after two hours' cautious work, in difficult rocky ground, the mist suddenly partially opened, and I found myself just above the north end of Llyn-y-Cwm. So I descended to the Pass amid falling waters and sheets of rain, and trudged down to Llanberis soaked to the skin, with my boots full of water. Dined at nine.

' 15*th*.—Out on the ridge of Glyder Fach tracing round the lines in the direction of the east side of Cwm Tryfan. Dreadfully wet. Yet I worked on in desperation, and as there were some intervals between the heavy storms of rain, I got a good deal done. Home by seven well soaked.

' 24*th*.—Out shortly after nine intending to have noted the section along the north side of the valley of the Llugwy. But in true geological fashion, I got led on and on to the top of Carnedd Llewelyn, and then taking advantage of the fine day, I walked all along the ridge to Carnedd Dafydd, and across Braich-du down to Llyn Ogwen. A glorious day and magnificent views of the Nant Francon range, with Snowdon at the back ; also all the country down to Cader, Aran Mowddy, etc. Home at seven.'

CAPEL CURIG, 30th August 1849.

MY DEAR AVELINE—I am in despair about getting away from here. With one clear day I could slash in a lot of country, all up as far as the watershed of Carnedd Llewelyn and Carnedd Dafydd, so that I am loath to leave to see you, lest that very day should occur when I am away Clear hill-tops are so scarce that one day when they are so is worth a fortnight of foggy weather. I have promised to make a run to Aber to look for lodgings for Jukes to occupy immediately after his marriage, and if possible I shall work my way there to-morrow, and next day trace a line from Bangor to Caernarvon, which would enable me to colour in a large piece of map, and so make the work look somewhat more forward. Early next week, then, I might perhaps manage to see you, for I am anxious to do so before going to Brummagem, where I act the swell groomsman to Jukes. It rains to-day without intermission.—Ever yours sincerely,

ANDW. C. RAMSAY.

Among the letters that came to him in this season of gloomy weather, the following note from De la Beche may be quoted :—

57 St. Stephen's Green, Dublin,
24th August 1849.

My dear Ramsay—Here I am once again. We had a famous passage last evening, and to-day I start, with Oldham, to the south.

If you go to the Wisdom Meeting [the meeting of the British Association at Birmingham], we can talk over some of our matters; and if not, I would get down to you afterwards. Matters are in good train at the Muzzy, and all going right, as it looks now; trumps will turn up there, I trust, next spring.

Rattling by the skirts of the Welsh hills last evening, the clouds seemed somewhat low, and looking up the valley of the Conway, I thought of the wet bother you have lately had, and of the troublesome quarters you are now in. 'Tis very tiresome for you. Once out of the high grounds of Wales, and we shall rapidly move ahead. —Very sincerely,　　　　　　　　　　H. T. De la Beche.

Ramsay did attend the 'Wisdom Meeting,' making a rapid journey thither, and acting with Jukes and Oldham as Secretaries of the Geological Section. But he was soon back at work again in North Wales. After carrying his boundary-lines from the Llanberis district northwards, until he had joined them up to those which had been mapped from Bangor, he left Llanberis on the 3rd September and stationed himself at Capel Curig, with the view of working out the structure of the group of mountains rising to the east of Nant Francon. Mr. Aveline was at work in the district lying to the north-east, and the two colleagues were enabled before the end of the season to join up their lines. Mr. Selwyn, having completed the survey of the ground lying between the Snowdon range on the north and Ffestiniog and Tremadoc on the south, was now at work in the Lleyn peninsula from Pwlheli. But there were still several portions of boundary to be settled along his northern limits. Ramsay had thus occasion to visit both his comrades from the central station of Capel Curig.

'*6th September.*—Attacked the side of Carnedd Dafydd; a hard day's work; was not home till half-past seven. I found the coffee-room full; Quakers in it who had been botanising.

'*8th.*—Started for Carnedd Llewelyn; glorious day and glorious day's work. Finished this side of the hill, all the way to the watershed, and was twice on the top.

'*2nd October.*—To Pen-y-gwryd. Struck up and had my last rap at old Glyder. I was sorry to part with him. Many a bright and many a stormy day have I passed on his sides, and as I scaled his cliffs many a happy hour have I spent *en route* home searching for ferns. The day was glorious, bright and warm. The world scarcely ever before seemed more bright and beautiful. I regained my voice and sang. I perfectly regained the use of my legs, and scaled the rocks strong and fearless as of yore.'

CAPEL CURIG, 26*th October* 1849.

MY DEAR AVELINE— . . . What precious weather since Monday till to-day! I got a good slash of work done to-day. In a few more days I must have a meet with you again to join up west of Llyn Crafnant and east of Llyn Geirionydd. I met Sir H. on Saturday at Bangor, and stayed with him till Monday. We had a short rap at Anglesey at very old rocks—older than the Cambrian.—Yours ever sincerely,

ANDW. C. RAMSAY.

The last sentence of this letter has a peculiar interest to geologists. It shows that the first impression made on Ramsay's mind by the older rocks of Anglesey was that they were pre-Cambrian. He afterwards came to regard them as altered Cambrian ; but his original and unbiassed judgment on the subject is now recognised to have been the true one.

CAPEL CURIG, 31*st October* 1849.

MY DEAR BILL[1] — . . . Winter does indeed approach, and it often looks sufficiently savage here, specially when the wind comes roaring down the glen, driving the rain before it in sheets for four whole days. Then ho! to see the rivers burst their bounds, and the lakes rise up a yard or two! Then old Kuhleborn reigns triumphant, and I, the enchanted knight, fall in love with all the female waiters and chambermaids, the daughters being lantern-jawed.

Then besides, I have work to do, and have begun to read up for the production of a third Introductory Lecture. What awful stuff the Wernerian disciples wrote, to be sure! I am busy analysing Jameson's (of Edinburgh) old writings. He was a disciple and pupil of Werner's, a favourite pupil, and by St. Anthony à Tours, I protest t' ye, it is about as easy to extract buttermilk from millstones, as to make sense out of the maze of words in which they lost themselves. And all that, too, under the guise of extreme exactitude!

But, somehow or other, o' nights, after a tough day in the air, I don't feel inclined for that dry work, or indeed for any serious work whatever. What then? Why, I have generally lots of letters to write, both of Survey import and in the friendly way. There's the home-circle, Sharpe, his honour Judge Johnes, Playfair's jewel, Mrs. Forbes, the Rev. W.

[1] His brother William.

R. S. Williams, our vicar, Dr. Falconer of Bath, and many others which (that I may not *now* weary myself writing lists of names, and so deprive my mother of the continuation of that inestimable catalogue with which, she will be glad to hear, I must fill my next letter) I forbear to mention. Then I now and then write verses. And yet again, when these delights fail, have I not some rare and delectable books, poets, and historiographers? For, look ye, how can a man weary with the choice and truth-telling histories of Alcofribas Nasier at his elbow, purchased by me at Birmingham for the small sum of 6s. 6d., and containing more wisdom and erudition than all the collected works of Hume and Smollet, Gibbon, Herodotus, Titus Plinius, Ferguson, Aristotle, Macaulay, Justinian (see his *Pandects*), Aulus Gellius, Avicenna, Froissart, Mrs. Trimmer, Bishop Stillingfleet, Machiavelli, Lamartine, Fox of ye Martyrology, Dean Swift, Phillip de Commines, Jean Paul Richter, Gawain Douglas, Knickerbocker, Anthony Count Hamilton, Barbour, the Rabbi Benjamin of Tudela, Ruddiman, Plutarchus, Mosheim, Mrs. Trollope, Thuanus, Rev. Thomas Burnett, Major Sabine, and many others, whose names I shall continue in my next letter?— Yours affectionately,   A. C. R.

'*2nd November.*—Magnificent day. Got a splendid day's work done, taking up the Llynbodgynwydd ash, and carrying it all round nearly to Trefriw, and so back by Llyn Geirionydd. It was a glorious day's work, and a glorious day to work in, so still and sunny. Speaking of peace, I conceived a sonnet on the way home, when I saw the mountains rise high and solemn into the sky in the twilight.'

Peace, vexĕd soul! there is a God above.
What though an evil destiny hath blighted
Thy fervent hope, quenching the dawning love
　　That, like a penetrating sunbeam, lighted
Life's shadowy path; beyond thy narrow care
　　The world is bright as ever.　Look around!
The earth is strewn with flowers, how passing fair!
　　The ringing voices of the brooks resound
In the low valleys, moss-grown rock and carn,
　　And the tall water-reeds reflected rest
On the deep bosom of the mountain tarn,
　　Telling of peace: the far-off mountain crest,
Piercing the sky, how strong, though tempest-riven!
Calleth aloud of rest, and points the way to heaven.

'19*th*.—A tremendous day's work with Selwyn, all across Dolwyddelan, up Cwm Penanmen, and round by Pwll Francon and Bettws y coed.'

CAPEL CURIG, 26*th November* 1849.

MY DEAR WILLIE[1]—The lines you allude to are Cowper's—

Would I describe a preacher, such as Paul,
I would describe him sober, grave, sincere, etc.

You will find them, I think, in 'The Sofa.'[2] It is a fine description. Martin Luther, however, is my favourite among more modern divines. A man also 'sober, grave, sincere'; but not always grave — a great divine and reformer and eke a great composer of sacred music, one who was not always grave, but sang his ballad 'with a full round mouth,' and was fond of a cask of good beer, as his letter to the Elector of Saxony (I think) proves, when he thanked him for one while attending a congress of divines. It is always worth living in the world while good beer

---

[1] His brother.
[2] The lines thus quoted from memory are not quite accurately given, and occur not in 'The Sofa,' but in 'The Timepiece,' line 395.

remains in it. We may thank our Saxon ancestors for that blessing.

' 28*th*.—After breakfast started for the hills above Llyn Bychan on the west side of the fault; finished them; re-mapped Mynydd Danlyn, and crossed to the other side by the lower end of Llyn Crafnant. While loitering about, taking a final look, I spied Aveline coming down anxiously, with his hat pulled over his eyes, his coat-collar turned up, his gaiters hanging about his heels, taking long strides and looking out ahead, but never holloaing, as another man might have done. So we joined and walked merrily down to Trefriw together.

' *4th December*.—Had a long consultation with Aveline and Jukes [at Aber] on the maps, and proved that Snowdon, Glyder, and all are not lower than the Bala lime and ashes. Jukes and I then started for the hills, and had a splendid day among the intrusive traps. Aveline returned to Trefriw, and Selwyn came up from Clynnog fawr. Joking and making fun all of us all night.

The campaign in Wales had thus lasted for fully six months, and was prolonged even into the stormy and inclement weather of December. It had been eminently successful, for a large tract of rough mountainous ground and complicated geology had been finished, and Ramsay had been able to join up the boundary-lines of his area with those of his colleagues on each side of him. And thus, turning his face southwards, and paying a short visit of inspection to Bristow in Dorsetshire, he was back in London before the end of December, to begin the indoor labours of another winter.

As before, we may take a few extracts from his diary of these winter months. The Geological Society continued to offer its fortnightly meeting as a rallying-point for the geologists in London. The Friday evening discourses of the Royal Institution, and the receptions of its genial Secretary thereafter, formed additional favourite gathering places. On the 1st March Murchison gave the discourse, and Ramsay records that this veteran geologist 'was quite nervous in the early part of his lecture, hesitating and leaving his sentences unfinished. But as he warmed he improved, and by and by got on very well.' A week later Edward Forbes occupied the same position, and his appearance is thus chronicled in the diary: 'The place was just about full. Forbes never appeared to such advantage. He lectured in first-rate style, coolly and boldly. The subject was "The Distribution of Fresh - water Fishes and Plants," which he treated certainly in a most masterly manner, showing that it depended on recent geological revolutions.' The next Friday is thus recorded: 'Royal Institution at night. The Astronomer-Royal lectured to a crowded audience, Prince Albert in the chair. Airy forgot himself, and lectured an hour and three-quarters! The Prince fell asleep.' The following Friday it was Ramsay's own turn to undergo the ordeal of addressing this critical and sometimes somnolent assembly. His account of the evening is as follows : 'I had half an hour's quietness in the little private room behind the theatre. At nine I was introduced, the Duke of Northumberland in the chair, the French Ambassador on his right, Mr. Hamilton on his left, and in the front row were Lord Overstone, Sir John and Lady Herschel, Wheatstone, Faraday, Murchison, etc. etc. It was literally a brilliant audience, with many ladies. The place was full, and they listened with great attention, occasionally quietly applauding, which gave me encouragement. I felt I was doing it easily. The praise I got from Herschel, Faraday, De la Beche, and others was almost too much to be good for me.' Faraday ran up to him at the close, shook him by both hands, and asked, 'Where *did* you learn to lecture?'

The subject of this discourse was 'The Geological Phenomena that have produced or modified the Scenery of North Wales.' The most interesting feature in it, considered with reference to the development of Ramsay's geological opinions, was undoubtedly the prominence now assigned by him to glacial action in connection with the landscapes of this country. This was the first occasion, so far as we know, when he made public profession of his belief in the former existence of glaciers in Wales, and gave at the same time new and original proofs of their presence, particularly instancing cases where mountain-lakes were still held back by ridges of terminal moraine, and where large blocks of rock were perched on ice-worn crags, where they must have been quietly deposited by the ice.

The annual festival of the Geological Survey took place on the 16th January 1850, and is thus recorded: 'Anniversary Survey dinner day. Sir Henry in the chair, Reeks vice. It passed off right jollily; lots of original songs from Forbes, Jukes, Baily, Smyth, Oldham, Hunt, Salter, and myself. I sang two.' One of his ditties was entitled the 'Song of the Geologues of the Woods,' and the concluding verse may be taken as a sample of its style :—

> The Survey needs no strangers,
> No scurvy council's bother;
> We'll work with Daddy De La Beche,
> And stick to one another;
> With six-inch sections, maps, reports,
> We yet shall see the day
> When Carlisle
> Shall blandly smile,
> And double all our pay,
> And every man shall keep his wife when he doubles all our pay.

The last day of April found Ramsay once more with Selwyn and Jukes at Merchlyn in North Wales. There were still various unfinished parts of his area to revisit and complete, likewise sundry lines regarding which he had to confer with his colleagues. The progress of the work rendered it necessary that some of the ground already surveyed should be gone over again in the light of fresh evidence. And after the surveys were completed there remained the laborious task of running horizontal sections across the area, including the most rugged and mountainous ground. These occupations, together with occasional visits of inspection, kept him busy in Wales until December— a long spell of field-work, only interrupted by a brief visit to London, the meeting of the British Association at Edinburgh, and an excursion to Dublin for the purpose of seeing his friend Oldham married. The life he led during those nine months is told in his diary and letters.

'*4th May*.—Out on the hills with Selwyn as far as the cliffs under Carnedd Llewelyn, and down by Melynllyn and Llyn dulyn. Got some good work done. Selwyn executed a most perilous feat of cliff-climbing; a slip and he would have been slain.

'15*th*.—Out again by Fawnog du and Carnedd Llewelyn. Its bald head was powdered with snow.

Yet the sun shone almost warmly, and having finished my work, I lay down on a big stone on Cefn-yr-Arrig and gazed on the deep shadows of Yr Elen and Ffynnon Caseg, the peaks of Carnedd Dafydd, and Y Glyder fawr, the great flats of Anglesey, and the distant outlines of Man and Ireland; and as I looked I felt my heart soften, and I arose a better man again. Came home over Y Foel Frâs, probably the last time I shall be on it.'

On the 18th of the same month he wrote to Aveline asking if he could recollect how many years the Survey had been at work in Glamorganshire, for, said he, 'in six weeks or so all this North Wales will be done, and I want, if possible, to compare times.'

On the 31st, in a letter to Salter from Caernarvon, he writes, 'Selwyn and I are here putting a final touch to all the difficulties and erst-seeming contradictions on this side the Straits. Marry, it comes out smoothly, except in so far that I fell on top o' the Rivals[1] yesterday, and so bruised my right shoulder that even writing is not a pleasant exercise for the arm. That is the beginning, I fear; what say you? Is it not terrible to think that now, when just finishing Wales, it is yet possible that I may this summer be found at the base of a cliff, with a bloody crown and my heels in the air?'

On the 6th June, while still revising with Selwyn from Caernarvon, he writes thus to Aveline : 'One long fine day will do for us here now, and a day or two's drawing. Then hey! for the sections. But first I purpose a run to Malvern for two days, to put in some alluvium left out by Phillips, and without which we can't publish that quarter sheet. I am a little

---

[1] Or Yr Eifl, a three-peaked hill (1887 feet) in the Lleyn peninsula.

bothered, but glad too, as I never saw the Malvern section.

'Our work here fearfully differs from Sir Henry's, and the worst of it is that he has, I think, published his opinion in his Anniversary address. It is about certain black slates which he puts *under* the Cambrian : they being, in fact, the Lingula (Silurian) beds brought against it again by a fault. It will be not a very agreeable job convincing him of this.'

The visit to Malvern and a hurried journey to London took up only some ten days, and by the 22nd June he was back once more at Llanberis to begin the arduous task of running sections. This operation was conducted with a theodolite and chain, the surveyor having the assistance of two men. The line of section having been determined in such wise as to cross the most instructive or important geological structures, and generally the loftiest summits, was drawn upon the map, and the surveyor then proceeded to measure on the ground the horizontal distances, and fix the relative heights of the various points along the selected line. These measurements were entered in a field-book, from which the section was afterwards plotted on a scale, vertical and horizontal, of six inches to a mile. When the outline of the ground had in this manner been correctly drawn, the geological structure was inserted from the maps and note-books, and, where needful, a final visit was made to the ground, and minor details were adjusted on the section. These operations, it may easily be believed, required both care and skill. They provided a further means of checking the accuracy of the maps, and when successfully completed, they furnished the surveyor with a valuable additional store of materials for the preparation of the written description of the geology of the district which he had mapped. How Ramsay fared with his sections across the Snowdon area he must be allowed to tell in his own words.

'*25th June.*—Out with my men to begin section from the top of Snowdon to the sea. Dodged the cliffs at the top, till from the Capel Curig road, attempting to make them chain back a bit to Pen Wyddfa, one of them refused, and I got exasperated, and discharged him on the spot. The fool was afraid to go over ground that I had danced over to show him the way ten minutes before. Home, annoyed at these Welsh blockheads.

'*26th.*—Got a new man and began, leaving the cliff till I had tried them. Came across over Craig du'r Arddu and found them more daring than myself; this will do.

'*28th.*—Out on the hills in a strong joyous mood. Did a tremendous day's work, chaining right along the face of the cliff from the top of Snowdon to the top of the Capel Curig path, and astonishing the sightseers by the strange peaky, cliffy places I planted myself on with my theodolite. Went to the top after, and took the angles of all the lakes and principal hills round. Home at seven. Went up Snowdon in an hour and a half, and down in an hour.

'*8th July.*—Out early. Carried on the section right down to the sea at Llanfair.'

The section-line that was now being traced ran on the one side from the top of Snowdon parallel with the Llanberis valley to the Menai Strait at Llanfair, whence it was afterwards continued across Anglesey. On the other side it was prolonged south-eastwards into the country mapped by Selwyn, and was carried by him into Merionethshire, across Cynicht, Moel Wyn, and Aran Mowddwy, and was continued by Aveline across Montgomeryshire. The plotting and final drawing of his part of the section occupied Ramsay's time in wet weather at Llanberis. The section, engraved by J. W. Lowry,[1] is one of the most striking in the whole series published by the Geological Survey. The geological structure is portrayed by Ramsay and Selwyn with a boldness and vigour, and at the same time with an artistic feeling, which had hardly been equalled in geological section-drawing.

The meeting of the British Association at Edinburgh offered a brief but pleasant break in these labours, as will be gathered from the following jottings.

' 1st *August.*—Murchison in the chair of Section C. Old Jameson[2] was there, and in the chair for a while. He looked just like a baked mummy. I spoke twice. We had some good papers; Forbes's first-rate, and Mr. Bryce[3] read a good paper as the mouthpiece of the Glasgow Natural History Society. What a rough, strong, clever-looking man Hugh Miller[4] is! My mother was there with Jess, looking very happy and venerable.

' 3rd.—This has been a glorious day. Went down to Granton at seven and embarked on board the *Pharos*

---

[1] An admirable engraver and accomplished mathematician, who engraved the Sections of the Survey for many years.

[2] Robert Jameson, born 1774, died 1854; the venerable Professor of Natural History in the University of Edinburgh. As mentioned previously (p. 155), he had been a favourite pupil of Werner at Freiberg, and for many years was the acknowledged leader of the Wernerian school in this country.

[3] James Bryce, born 1806, died 1877; one of the masters of the Glasgow High School, and author of some geological papers and a little volume on the geology of Arran and Clydesdale.

[4] Hugh Miller, born 1802, died 1856; originally a stonemason in the north of Scotland, devoted his attention to the Old Red Sandstone of that region, and afterwards wrote some excellent popular books on the fishes of that formation, some of which he was the first to discover.

steam-yacht, belonging to the Commissioners of Northern Lights. Dr. Robinson, Strickland, Dr. Johnston of Durham, Oldham, Allan, M'William, Williamson, and many others there,—a most lively and amusing party. We got into boats by and by at the Bell Rock [Lighthouse], and fairly effected a landing. A wonderful sight that tower, rising direct from the waters, so far away at sea! Then we went to the Isle of May and the Bass Rock, where we landed and saw its wonderful covering of live birds. There we picked up Lord Wrottesley [1] and his daughter. Then to Inchkeith, and so home. We breakfasted, lunched, dined, and had tea on board, and gorgeous meals they were. Some splendid speeches were made, and altogether it was quite an event in one's life. Strickland had a gannet knocked down with a hammer, to take away with him. In the evening to Robert Chambers's : a large assemblage.

' *6th.*—Breakfasted at Chambers's. Sopwith very funny ; he is witty. Opened the Section by giving a very short abstract of my paper. Sedgwick and Murchison then spoke of the labours of the Survey. I spent the rest of the day at the Ethnographical Section. Latham spoke a splendid paper to the few gentlemen round the table, Mrs. Latham and I frequently making the whole audience. Went to the soirée in the Music Hall. When just over, Forbes and I, to Sir David Brewster's great disgust, got up a dance in the Assembly Rooms. We had nice little partners, but neither of us knew their names.'

After the close of the Association meeting he spent a few days in Glasgow with the old familiar faces. One little touch may be quoted from his diary : ' Then the parting. My mother came upstairs.

---

[1] Afterwards (1854) President of the Royal Society.

"Come back as soon as you can, for you'll not have to come often now," she said, and I was obliged to break away and retire to my own room for a little.' By the 16th August he was back once more at Llanberis, whence he transferred himself to Bethesda, in order to get at various outlying pieces of ground around Carnedd Dafydd that remained still incompletely surveyed.

De la Beche, who was never happier than when he was able to report the completion of a large number of square miles, began to be fidgety about the length of time taken by the section-work in Wales, and the consequent diminution of the area of ground surveyed. Ramsay complained to Aveline on the 27th August that it was unfortunate to be carrying on this work 'against the grain with the governor, for he would fain take us away and leave the thing unfinished. I shall get away by the middle of September. You will not get off so soon, I suppose. About a week ought to finish my mapping out of doors. Two days indoors or three, some bad weather (as to-day), and a diabolical section from Bettws over Pen Llithrig - y - Wrach, Carnedd Llewelyn, and the sea—the thing is done.'

' *5th September.*—Out by Carnedd Dafydd, tracing in the drift. Got a good many wrinkles on the subject. It must have been 2000 feet high at least. Came down on the Carn Llafa side of Carnedd Dafydd and corrected these alternations by means of the faults —a most troublesome bit of work. Home at half-past six.

' *11th.*—Did a glorious day's work with Howell[1] up

---

[1] H. H. Howell, who joined the Survey in 1850, became District Surveyor in 1872, Director for Scotland in 1882, and Senior Director in 1888, an office which he still holds.

as far as Aber, getting all Jukes's ugly bits of sandstone, etc., perfectly explained—a succession of domes
cut off by faults.   Home at half-past seven—a long,
long walk.

'12*th*.—Up by the coach to Cwm Idwal.   At
the top we found a splendid haul of fossils, and I
made a grand discovery respecting the drift.   [He
here gives the section across Llyn Idwal, afterwards
published in *Quart. Journ. Geol. Soc.* vol. viii. (1852),
p. 375, showing the drift capping the summits above
the lake and a moraine forming the barrier of the
water].   The moraines of these valleys are subsequent to the drift, because, if previous, they would
have been smothered in it.   But, as I before proved,
the *roches moutonnées* are previous to it, because they
are covered by it up to great heights.   The drift on
top of Cwm Idwal is 2500 feet high, and it reaches
probably a parallel height on Cwm Llafar, being
thence connected all the way with the drift of the
sea-side.'

By the end of this month he was at Dolgelli,
helping Selwyn to put some finishing touches to the
mapping of the Cader Idris region.   On the 12th
October he was able to make to Salter an important
announcement touching the troublesome regulation
as to receipts for travelling charges.   'Henceforth
and for ever you take no more receipts for travelling
expenses, and in place thereof you must make out
a travelling charges bill.   I've got it all in right order,
and by a magnificent stroke of genius have got Sir
Henry's formal consent thereto.'

His colleague Oldham had determined to resign
the charge of the Irish Geological Survey, and to
accept the direction of the Geological Survey of

THOMAS OLDHAM

India.[1] As a preliminary step he arranged to be married, and asked Ramsay to support him on the wedding day as groomsman. So the Welsh work was laid aside for a week, and Ramsay for the first time went to Ireland. He says of his reception at the house of the bridegroom that he was formally introduced to the family, including ' Mr. Neptune Oldham, a big Newfoundland dog, who was sitting on a chair at table, finally shaking hands with the dog, who presented me with his paw in the most courteous manner. We all got at home with each other at once.' One after another of his colleagues was thus quitting the ranks of bachelorhood, and he could not help heaving a sigh now and then, and wondering if his own time were ever to come. Writing to Oldham a day or two after the marriage, these feelings escaped in verse :—

> Thomas hath found what he desired,
>   The maid his heart did fix on ;
> He by an angel was inspired
>   When he popped to Miss Dixon.
>
> Another bachelor hath passed,
>   And I, for lack of gold, boys,
> Ah, woe is me ! am falling fast
>   Into the vale of old, boys.
>
> Oh, many a sheep's eye have I thrown,
>   Have cast full many a lamb's eye,
> But never yet have chanced on one
>   That cared to take a Rams-eye.
>
> Would that the gods might yet be kind,
>   Nor longer try their tricks on ;
> Then haply even I might find
>   Just something like Miss Dixon.

---

[1] On his resignation he was succeeded by J. B. Jukes, who, having joined the service in 1846, was transferred from the English Survey, and became Director of the Geological Survey of Ireland in 1850, retaining that post till his death in 1869.

The fascination of glacial geology was now at length beginning to influence Ramsay's geological bent and to tinge all his views of Welsh scenery. He had practically finished the survey of the solid rocks. Their problems, though by no means all solved, had at least been so far settled as to allow of the preparation of maps and sections for the engraver. The compilation of the descriptive memoir of the region would be a laborious task, involving years of interrupted application, and many renewed visits to the ground. But the glaciation of these Welsh mountains had all the charm of novelty. Buckland, Darwin, and others had described some of the proofs of former glaciers, but no one had yet attempted to trace the story of the successive changes of geography and of climate recorded in the various glacial deposits. We now find in Ramsay's note-books and diaries frequent reference to the subject. While stationed at Bethesda he made numerous observations and compiled many notes relating to the ice-markings on the rocks, the distribution of the drift, the grouping of perched blocks, and the position and heights of moraines. He was in this way gradually accumulating materials for his first essay on the glacial phenomena of this country which he communicated a year later to the Geological Society.

There still remained a portion of Anglesey to be surveyed before the maps of North Wales could be regarded as complete and ready to be prepared for the engraver. De la Beche had himself traced the lines across some parts of that county, and other portions had been mapped by W. W. Smyth. Ramsay and Selwyn early in November crossed into Anglesey with the object of filling in the unfinished portions and completing the whole. The following letter from De la Beche gave them his impressions of the structure of the ground immediately after they had begun their work :—

LONDON, 11*th November* 1850.

MY DEAR RAMSAY—Touching the mica-slates, chlorite-slates, and other matters of the lower ground in Anglesey, they are, of course, what they can be proved to be; and no matter what they may be, let us get at the fact. Pray keep a bright look-out for the conglomerates; they are most valuable in such investigations. You have probably examined that beneath so much of the Cambrians as is to be seen on the banks of the Menai, near Bangor. The conglomerates nearer Llanberis show clearly that the matter of the Cambrians there is, in part at least, compounded of older detrita rocks—kinds of quartz-rock being among them. If it be really right that the Bangor beds are these said affairs brought up again, probably similar pebbles will present themselves. Here, then, we have evidence of detrital beds consolidated before so much of the Cambrians as such conglomerates may form the base of.

I know not how you have attacked the ground, but if I had been with you, which I very much regret is not the case (there are, however, matters of more pressing importance now under consideration here), I should have made you master of the country at Holyhead Island, and have proceeded across country to Amlwch, though not quite direct. Taking up the black shales (graptolitic) based upon conglomerates of variable character, but sometimes containing pieces as large as one's head, to these succeed a parcel of trappean affairs—limestones beautifully laminated; above these, shales and more arenaceous rocks, sometimes purplish, and so on to the northern coast, where heavy conglomerates with some impure limestones cover all. A better section is no doubt to be obtained on the sea-coast by means of a boat, but such means of conveyance are now (November) out of the question. The two sections confirm each other, some beautiful granite veins and alterations near them requiring a little caution.

At Amlwch the sections are capital, on the coast especially; the Parys mine, a continuation of the graptolitic slates. Near the place with an unpronounceable name, to which I direct thee, there are some capital conglomerates. Pray look the pebbles well over. Henslow called these Old Red; they are not so.

The upper purple beds occupy a position very like similar beds in Ireland—the highest of the series there known to us in Wicklow, Wexford, and Waterford. The date of the granitic intrusions of Anglesey is clearly that of the Irish country noticed—anterior to the deposit of the Old Red Sandstone. These upper purple beds will interest you, not that there is anything in purple and (their common associates) greenish beds; they are found of all ages. The upper purple beds in Ireland, the position of which is undoubted, often remind one of the Cambrians of Bray Head and other places.

It seems to me, before anything be written or published, it will be needful for you and self to go over some of the main sections and points. This you will be the better able to do after your present examinations.

I have not the maps with me; indeed I am writing away from the Museum, and therefore cannot point out more distinctly where I would wish you to look. There are some capital cases of smashing on the coast from granitic intrusions beyond (southward of) the range of the rocks holding the Parys mountain mines—really good things; so is the whole coast. I believe I have walked or boated the whole in Anglesey. I should like a run with you in Anglesey, and, please the small porcines, we will have one, whether the lower rocks be Tertiaries turned topsy-turvy or superfine elders. I am called to attend to other things.—Yours sincerely,

H. T. DE LA BECHE.

It will be obvious from this letter that the Director-General had recognised conglomerates at the base of the Cambrian series of Anglesey, that he wished to keep an open mind as to the relations and age of the rocks underlying these conglomerates (which he seems to have been inclined to class as pre-Cambrian), and that he had observed the presence of trappean or volcanic intercalations among the older Palæozoic formations of the island. It was unfortunate that on all these points, where he was undoubtedly right, his able lieutenant came to differ from him. Selwyn, indeed, clearly detected the unconformability of the lowest Cambrian strata upon an older series of schists. But on the maps as finally published Ramsay's views prevailed. No pre-Cambrian rocks were there shown. The crystalline schists were classed as 'altered Cambrian,' and the existence of volcanic breccias and other proofs of volcanic action were not recognised.[1]

[1] See a discussion of this subject in Presidential Address to Geological Society, *Quart. Journ. Geol. Soc.* vol. xlvii. p. 81, 1891.

Apart from the geological work, there is a peculiar interest in these few weeks of Survey doings in Anglesey, for now, unconsciously, Ramsay was approaching one of the momentous epochs of his life. During the day he and Selwyn traversed the rocky northern coast of the island, charmed with 'the cliffy foregrounds, the white breakers, the great misty plains of Anglesey, and the snow-covered mountains rising beyond so still and grandly.' At night they had the shelter of little inns, sometimes of the homeliest kind. In the course of their traverses they received an invitation to make, for a day or two, the rectory of Llanfairynghornwy their headquarters. The following notes from his diary convey Ramsay's first impressions of this hospitable household: 'The house is somewhat characteristic, being full of all sorts of odds and ends, and not in the highest order, yet everything telling that they are people who do not exclusively busy themselves with externals. There is a character about the family. Mr. Williams is one of the best specimens of a Welsh clergyman I have met, polished and conversational, not at all deep, but very agreeable, and, I should say, conscientious and hard-working. Mrs. Williams is a remarkable woman. She was engaged [when the two geologists arrived] enlarging a map of Palestine for the use of a school her daughter takes care of. They all assist at wrecks, etc., and she has made a survey of the Skerries, taking the angles with a prismatic compass. They [afterwards] made me explain the glacial theory, and were, I think, interested, especially Miss Louisa, who is certainly a very clever girl.'

The geologists were asked to come back and spend Christmas at the rectory. This pleasant visit is thus referred to in a letter to William Ramsay, written from Llanfairynghornwy on Christmas Day: 'We were detained at Bangor at work till the last moment, and when done we threw ourselves into the rail, and fled away here yesterday evening to eat a Christmas pie with our jolly friends the Williamses, and eke a goose with apple sauce. Marry, come up; I'll stay a day or two and make myself merry when I am here, for we've been working extra hard. They (the W'ms.) are bricks, and no mistake. It is no joke to enter into a contention with one of the young ladies, Miss Louisa; she is so witty that you might just as well cut your eye-teeth before you begin.' From the very first he was greatly interested in this bright, clever daughter of the house. In his diary he makes frequent reference to her: 'Wit and a sense of the ludicrous is her characteristic; sense she has a good deal of, and warm-heartedness no end of.' 'Commenced the year (1851) dancing a polka in the hotel ball-room, Chester. Trifling and merry enough, I believe, with the witty Louisa for a partner; not ominous, I opine, of future partnership.' Whether 'ominous' or not, the acquaintance developed into sincere affection on both sides, and he found here at last the loving and devoted woman who a year and a half afterwards was to become his wife. But these pleasant dissipations, so fitly closing a long and arduous season of field-work, soon came to an end; and by the 5th of January Ramsay was once more at his post in the Survey Office in London.

The building in Jermyn Street was now rapidly approaching completion. The collections at Craig's Court were being transferred to their new home. Already the offices of the Survey had been removed.

There was, therefore, all the bustle of preparation in the staff. Moreover, Sir Henry's great scheme for the foundation of a school of applied science seemed now at last almost certain to be carried out, and if so, it would involve considerable change in the positions, duties, and emoluments of a number of the officers of the establishment. Add to this that the Great Exhibition of 1851, which would open in a few months, was the subject of much consideration in several Government departments, and not least among the officers of the Museum of Practical Geology. Occasionally a minister would come to inspect progress. Prince Albert himself went carefully over the building and its contents, and took much interest in it. Among the official visits there was one which is thus narrated in the diary. '*6th March.*—Lord and Lady John Russell and two children came here to-day. He, cold and uninterested; she, most charming and intelligent. When I was introduced, he merely bowed coldly. Ditto to all. Blewitt, the M.P. for Monmouth, he coldly bowed to. "Who would have thought," said Blewitt, "that I've sat beside that man and supported him for fourteen years; he is a nice man to keep a party together!" I had a good deal of conversation with Lady Russell, and was much pleased with her.'

The Anniversary gathering of the Survey this winter was the most successful that had yet been held. It is thus recorded: '18*th January* 1851.—Busy at the Museum till nearly half-past five. Then off for a short walk, and so to the Imperial Hotel, Covent Garden, to the Annual dinner of the Royal Hammerers. And oh, wasn't it a jolly dinner! We were: Sir Henry, Forbes, Captain James, Captain Ibbetson, Smyth, Aveline, Bone, Baily, Bristow, Salter, Reeks, Selwyn,

James Forbes, Playfair, J. Arthur Phillips, Hunt, Jukes, Oldham, and myself. Oldham sat on Sir H.'s right, and I beside him. After dinner the mirth became fast and continuous. One comical song followed another, all original. Forbes made me roar with laughter, chanting something at me about—

> ' I'll lay my head on a Bala Bala bed,
>   And wed a parson's daughter.

My songs were, one to the tune of "Trab, Trab" (trap-trap, rap-rap, map-map), and the other, "O weel may the Survey speed," etc.' A verse or two of the second song, which was headed ' 1841-1851,' may be quoted :—

> I joined the chief in Tenby Bay,
>   And shillings I caught nine,
> 'Twas three for breeks, and three for beer
>   And shillings three to dine.

> When first I left the Land o' Cakes
>   And took to wearing breeches,
> I little thocht that I should join
>   This corps o' De la Beche's.

> There's Forbes's men that work within,
>   And our field-working laddies,
> Including Jukes, that shaved his chin
>   To please the Irish paddies.[1]

> When age has put our auld pipes out,
>   By precept and harangue,
> New lads will rise without a doubt,
>   Will gar the hammers bang.

---

[1] Jukes wore a copious black beard, but when, on obtaining the Directorship of the Geological Survey of Ireland, he became a candidate for the Chair of Geology at Trinity College, Dublin, he was informed that this formidable facial appendage would probably create a prejudice against him in the University circle, so he appeared one morning at the Holyhead hotel with his chin clean shaven, and so altered in appearance that the waiters took him for a stranger. But, as he did not get the professorship, he allowed the beard to grow again, and it soon became as exuberant as before.—See Jukes's *Letters*, p. 454.

The Anniversary dinner of the Geological Society was this year chiefly memorable for one of the most wonderful exhibitions of Sedgwick's oratory. 'At the dinner,' says Ramsay, 'Forbes, Wilson, Aveline, Smyth, Sopwith, Captain James, Logan, and a few more of us got together. Hopkins, the new President, was in the chair. He was slow. Sedgwick made the great speech of the evening. By turns he made us cry and roar with laughter, as he willed. His pathos and his wit were equally admirable. Home at twelve.'

To the Geological Society Ramsay communicated this winter his first paper on glacial phenomena. For nearly three years he had been giving increased attention to this subject. Not only had he met with many new illustrations of the history of the glacial period, but his observations, now that his eyes were opened to the existence and significance of the facts, led him to perceive the meaning of many scattered surface-features in South Wales, to which, at the time he was surveying in that region, he had paid little heed. His paper was read on the 26th March 1851, and was entitled, 'On the Sequence of Events during the Pleistocene Period as evinced by the Superficial Accumulations and Surface-markings of North Wales.' His comment on the meeting of the Society runs as follows : 'Read my paper at the Society. No man objected but Hopkins, who said little, however, being President, and he only objected to one point, and praised all the rest. Sir H. made a capital speech, and I think made an impression on Hopkins on that very point that bothered him in my paper. Murchison, Lyell, and the rest scarce ventured to criticise my views, though they spoke well for the grasp and importance of the paper.'

N

A week after the reading of this essay the following entry occurs in the diary.   'Went over my Welsh
glacier-maps at night.   Walked up each valley with
my mind's feet, and took Logan with me.   He said at
the close that he thought I had proved my case, but
that before publication I had better look at a few
points again.'   Whether it was this advice of the
veteran Canadian geologist, or the criticism at the
Society, or his own mature reflection that determined
him, he withheld the publication of the paper for more
than a year, and then issued it with a slightly altered
title.[1]   The chief point insisted on in the paper was
the fact that the so-called glacial period embraced two
distinct glaciations : one widespread and prior to the
deposition of the Drift ; the other local in valleys and
later than the Drift.

A subsequent meeting of the Society is thus described : 'Murchison had a paper on the Denudation
and Drift of the Weald of Sussex.   When the debat-
ing came, Lyell first spoke indifferently, unable to
overcome the difficulties, but evidently feeling that
Murchison's catastrophic solutions were the greatest
difficulties of all.   Then followed Sharpe, who said that
one would suppose from M.'s reasoning that elephants
were marine, instead of terrestrial animals.   Then came
Mantell, who, in a most eloquent speech, asked, if the
great mammifers were annihilated by this catastrophe,
how is it that their bones are always found scattered
and in fragments ?   Would not the ligaments and skin
keep them at least so far together that we would find
the principal parts of the skeleton near?   Then
followed Forbes on the same tack, then Dr. Fitton,

---

[1] 'On the Superficial Accumulations and Surface-markings of North Wales,'
Quart. Journ. Geol. Soc. viii. (1852), p. 371.

asking for more facts and less theory, and then myself, showing how little dependence was to be placed on angularity or non-angularity of pebbles as a test of date. Every one came hard down upon him. . . . He thought he was to be received with praise, and every one opposed him.'

The Red Lions had a curious experience this season, of which the diary contains the subjoined account. 'At six went down with Forbes to the Red Lions at Soyer's. It appeared that he had a great dinner to the Press, etc., of all nations, and having made no provision for us, he dodged us into dining with them in the great hall. His first request was that we should dine at the same hour to save his cooks. There were all the Reds of note, including Owen, Latham, Dr. Smith, etc. etc. He appointed the best places at table for us, and made his people ply us with all sorts of good dishes and wine. It was a splendid joke. In the garden was a huge oven, in which half an ox was roasted. At a signal the covers were removed, and it was wheeled on to great dishes on a hand-barrow. Twelve cooks carried it, and a brass band marched before playing "The Roast Beef of Old England," while all the guests came up behind laughing. The Honourable Captain Fitzmaurice, Soyer had secured as principal toast-giver and speech-maker. This man had indicted him [the great French cook] as a nuisance, with his lights and bands o' nights. Soyer called thrice, but the Captain would not see him; at length he somehow forced himself into his presence, and lo! the gallant Captain now sat by his side, and returned thanks for the Army and Navy. The whole thing was so cleverly done that, save Latham, perhaps, all of us took it as a joke and

laughed prodigiously.   Before dinner, when some of us
looked a little displeased, and Ibbetson and Henfrey
remonstrated, Soyer looked round for the meekest
man, and seizing Van Voorst, "Come," said he, "let us
talk it over," and marched him away arm in arm.

'It was glorious to hear Jules Jamin reply for the
press, so rich was he in French grimace.   Forbes I
spirited up to reply for the Lions, which he did in a
great row, but with great humour.'

In spite of the multifarious London duties of this
winter and spring, Ramsay contrived to secure a few
days in the field, inspecting some of the joint work of
Forbes and Bristow in the Isle of Wight and along
the Dorsetshire coast.   Of this pleasant but brief
Easter excursion he records as follows :—

'*Easter Monday.* — At the railway-station met
Lyell and Bristow.   Forbes met us at Southampton,
and so, by way of Lymington and Yarmouth, we got
to Freshwater Gate by half-past six, and dined at
half-past seven.   I liked Lyell better; he was often
anecdotical, but principally geological all day.   He
laughed tremendously when Bristow said his portmanteau was so heavy because it contained De la Beche's
new "Geological Observer." [1]

'*25th April.*—Spent the whole day at Warbarrow.
Forbes has certainly made a capital story of his divi-
sions of the Purbecks, which we must follow if
possible.   We saw a splendid section all along the
coast from thence to Kimeridge Bay, where we got at
five, and came back in the fly.

'We all like Lyell much.   He is anxious for

---

[1] The first edition of this portly volume, not being divided into chapters,
was a formidable piece of reading, more especially as Sir Henry's style was not
always of the clearest.   The book was sometimes irreverently called by outsiders
'The Jermyn Street Bible.'

instruction, and so far from affecting the big-wig, is not afraid to learn anything from any one. The notes he takes are amazing ; many a one he has had from me to-day. He is very helpless in the field without people to point things out to him ; quite inexperienced and unable to see his way either physically or geologically. He could not map a mile, but understands all when explained, and speculates thereon well.' 'He wore spectacles half the day, and looked ten years older [in consequence]. Logan says it is vanity that prevents his always doing so. I think it is custom, and perhaps his wife.'

CHAPTER VI

THE scheme which De la Beche had so patiently worked at for some twenty years was now at last brought to its consummation. He had succeeded in inducing the Government to build a spacious edifice, extending from Piccadilly to Jermyn Street, which was to be entirely devoted to the purposes of Geology and its allied sciences. The main portion of the building was arranged for the display of specimens of minerals, rocks, and fossils, especially to illustrate the geological formations and mineral products of the British Isles. A large series of admirable specimens had been obtained from the mines of Cornwall and Devon, showing the characters of metalliferous veins and their accompaniments. Another series represented various mineral substances employed in manufactures or arts, with examples of their successive stages of treatment from the raw material to the finished article. A third series consisted of various stones employed in building or for decorative purposes. There were likewise numerous specimens of mining tools and machinery and models of mines and pit-workings. In every way that could be devised the contents were so chosen and arranged as to justify the name given to the building, 'The Museum of Practical Geology.' The old collec-

tions at Craig's Court not only found now a worthier
domicile, but they were augmented by many specimens, which, for want of room, could not previously
be exhibited.

But in keeping the practical application of geology
before the eyes of the public, the claims of pure
science were not lost sight of or held in the background. A fine assemblage of fossils which the
Geological Survey had gradually been amassing was
now arranged in due stratigraphical order. The
visitor in walking round the galleries had before him
the characteristic plants and animals of each great
period of geological time, all properly named and
grouped. He could thus, text-book in hand, study
the fauna and flora of any particular geological period
with a fulness and ease never before attainable. Geological maps of different parts of Britain were
suspended for reference. Every effort was thus made
to ensure that for purposes of serious study the
Museum should be as useful as possible. By this
combination of the systematic and the practical it
was believed that an important step was taken in the
development of geological science.

But the Jermyn Street Museum only carried out
more fully and with ampler space what had been
already attempted in the more restricted quarters of
Craig's Court. Though the giving of lectures in connection with the Museum had been sanctioned as far
back as 1839, the want of proper accommodation had
prevented this design from ever being put into execution. But there was now the possibility of better
things, and the great new departure in the organisation was the creation of a special teaching staff and
the establishment of a definite curriculum of scientific

training.   Other countries had long had their schools
of mines, yet Britain, with its enormous mineral
wealth, then yielding twenty-four millions of pounds
annually, had never possessed such an establishment.   It was known that vast sums of money had
been wasted in fruitless search for minerals, where a
knowledge of geology would have shown that such
minerals did not exist.   It was admitted that science,
if consulted in such cases, could direct the search for
minerals in new localities, and aid in the proper and
economical working of those already known.   Many
representations had been made to the ruling authorities of the country, urging the great need of scientific
instruction in all branches of science capable of
assisting in the development of the mineral industries
of Britain.   But it was not until the early summer of
1851 that the idea was finally launched into practical
accomplishment.

The claims of De la Beche as the originator and
the life and spirit of this comprehensive scheme were
never more forcibly urged than by Murchison when,
four years later, the Geological Society awarded its
Wollaston Medal to the Director-General of the
Geological Survey.   'Then arose,' he said, 'and very
much after the design of the accomplished Director
himself, that well-adapted edifice in Jermyn Street,
which, to the imperishable credit of its author, stands
forth as the first palace ever raised from the ground in
Britain which is entirely devoted to the advancement
of science! . . .   It is our bounden duty [as members of
the Geological Society] to cleave closely to our off-
spring, Her Majesty's Geological Museum — nay
more, to use our most strenuous endeavours to have it
maintained by the British Government in that lofty posi-

tion to which it has been raised. We must, in short, not only hold firmly to, but act upon the faith which is in us, and see that an establishment like this, though it naturally branches off into highly useful and collateral subjects of art, be never rendered subsidiary to them, but be permanently and independently sustained on its own solid basis of pure *science*. This, our view, will also be taken, I feel confident, by every enlightened statesman who may be placed in a station to provide for the future well-being of the admirable Museum, founded and completed by our Wollaston Medallist.'[1]

The 12th May was fixed for the formal opening of the Museum by Prince Albert. Ramsay thus chronicles the events of the day: 'Over [to the Museum] soon, wound up all I had to do, and then prepared for our opening. Crowds began to assemble about half-past eleven. I helped to receive below. By and by the Prince came. We of the Museum, some of the ministers, etc., sundry Lords once of the Woods, the Bishop of Oxford, some of the geologists, etc., followed to the vacant chair. Sir H. read an address, the Prince read a reply. Then we all walked round, Sir H. leading, and each officer explaining his own department. And it was over.

'A terrible damper occurred which we kept from Sir H. Faraday told Hunt just before the Prince came that poor old Mr. Richard Phillips had died yesterday. It was a shock to me. Strange that he should have died just at the opening of the Museum. I find myself unconsciously repeating his jokes. We shall see him no more toddling about with a joke for every one.'

As finally adjusted, the subjects to be taught at the

------

[1] *Quart. Journ. Geol. Soc.* xi. (1855), pp. 24, 25.

newly-instituted 'Government School of Mines, and
of Science applied to the Arts,' and the officers by
whom the courses of instruction were to be given, were
as follows :—

President—Sir Henry T. De la Beche, C.B., F.R.S.

Chemistry, applied to the Arts and Agriculture—Lyon Playfair,
Ph.D., F.R.S.

Natural History, applied to Geology and the Arts—Edward
Forbes, F.R.S.

Mechanical Science, with its Applications to Mining—Robert
Hunt, *Keeper of Mining Records.*

Metallurgy, with its Special Applications—John Percy, M.D.,
F.R.S.

Geology, and its Practical Applications—A. C. Ramsay, F.R.S.

Mining and Mineralogy—Warington W. Smyth, M.A., F.G.S.

His acceptance of the lectureship of geology in
this institution rendered it necessary that Ramsay
should vacate his chair at University College. On
the 15th June he sent in his formal letter of resignation. There was a disposition on the part of some of
the College authorities not to continue the professorship
after he should give it up, but to send the students to
him at the School of Mines. He himself, however,
was adverse to this proposal, and the idea was
abandoned. The teachers of the School did not
aspire to be called 'Professors,' and Smyth used
almost angrily to resent the appellation. But Ramsay
having for four years worn the gown in a chartered
college, the name of Professor continued to be given
to him, in accordance with the northern proverb, 'Once
a bailie, aye a bailie.'

The preparation of lectures for the new school was
a much less arduous task than that which presented
itself to him four years before. The course he had
given at University College would suffice for his

WARINGTON W. SMYTH

purpose.   He had not written out his lectures, but had only made full notes, and these he used to revise frequently, so as to bring them abreast of the onward march of geology.   This task had to be accomplished before the beginning of the next year.   But it was not one which pressed heavily on him, even though it included the preparation of a special introductory lecture designed for the purposes of the School of Mines.

Ramsay had thus ample time for inspecting duty in the field during the summer and autumn.   Much of the earlier part of the season was spent in the Midlands looking over the ground mapped by or assigned to Jukes, H. H. Howell, and E. Hull.[1]   The two latter geologists were recent additions to the staff, and he trained them for their work.   Never was there a more delightful field-instructor than he.   Full of enthusiasm for the work, quick of eye to detect fragments of evidence, and swift to perceive their importance for purposes of mapping, he carried the beginner on with him, and imbued him with some share of his own ardent and buoyant nature.   Laziness and indifference were in his eyes such crimes that indulgence in them marked a man out for his wrathful indignation, and even for ultimate dismissal from the service.   He would take infinite pains to make any method of procedure clear, and was long-suffering and tender where he saw that the difficulties of the learner arose from no want of earnest effort to comprehend.   But woe to the luckless wight who showed stupidity, inattention, or carelessness !   Ramsay's eye would flash, his hand would whisk the tips of the curls on

[1] Edward Hull joined the staff in 1850, became District Surveyor for Scotland in 1867, was appointed Director for Ireland in 1869 on the death of Jukes, which post he held until his retirement from the service in 1890.

his head, he would seize the map and rush ahead, calling on the defaulter to come on and look. And he would keep up this offended tone until he felt that his pupil had at last been made to feel his delinquency. Then some snatch of a song or line of an old ballad or fragment from Shakespeare, appropriate to some phase of the incident, would come into his head, and instantly it would be on his lips with probably a hearty laugh, that showed how entirely the cloud had passed away. If a man had any geological faculty in him, it was impossible that it should not be stimulated and educated under such a teacher. And if, unhappily, there was no such faculty, Ramsay soon discovered the defect, and after full trial the recruit was advised to seek other fields of exertion.

The inspecting duty in the Midland region brought Ramsay into close familiarity with a type of English scenery which contrasted strongly with what, during his Survey life, he had been chiefly used to in Wales. Thus he writes : ' 19*th July.*—Up into that fine wild part of old England by Cannock Chase. It truly gives an idea of what much of England must have been in the days of Robin Hood—wild, undulating, unenclosed ground, covered with heath and bracken, and here and there sprinkled with oaks, birches, and alders. In the woods and on the hillsides you may see the wild deer trooping along, while now and then you raise a lazy heron, or the whirring grouse and black game.'

' 23*rd.*—Out to Maxstoke Priory, etc. [Warwickshire], tracing on Howell's fault. What a noble place that has been, with its piles of building, its great cathedral-like church, and its perfectly-built encircling close walls of smoothed stones with buttress and sloping copings ! I was charmed and grieved at the sight of the stately ruins; scarcely anything remaining but part of the great church-tower, the gateway, some of the smaller buildings, now a farm-house, and these beautiful walls. To-night I heard the Shakespearian word "pudder" used for the first time in conversation. Old Mr. Brown of the Colesleys said, "It will be a fine day to-morrow, if the thunder does not pudder up," pronouncing the *dd* as *th*. It tells a singular story to see many of the old farms surrounded by moats in these parts.'

The weather during part of the time in Derbyshire was excessively warm, and made field-work somewhat trying, as the following characteristic letter will show :—

ASHBOURNE, DERBYSHIRE,
30th June 1851.

MY DEAR SALTER—Where you may be I know not, whether above or below ground, recent or fossil. . . . Here we are burned up with fervent heat, and our souls are melted within us. Ginger-beer o' days is the only drink, and we dine at twelve o'clock at night with bitter beer and soda-water. Our noses are flames of fire, and our lips breathe smoke as a furnace. Oh for the dim cellars of the Museum, and a pint of cool stout with an oyster! Then should our throats be opened, and our lungs sing aloud like a game-cock. Hip-hip-hurrah for Lord S——, who is not quite so bad as he's ugly. With a shout for Sir Henry, the Gov'nor, and a prayer that his legs may grow stouter; Stout as the legs of strong Samson, who bore off the gates of a city, Easy as Salter would carry a trayful of shells oolitic Up the high gallery - stairs, where calamites ever

reposing, Rest in their timber-glass tombs, delighting
the eyes of the public; Telling a tale of past epochs,
a tale of the forests primeval, When mighty batrachians
crawled o'er the mud that encircled their rootlets, And
the convex *Productus* clung by byssus to stem and to
stump, sir; Like to the oysters that stick to the
mangroves afar by the Quorra.—Quoth
                                        ANDW. C. RAMSAY.

The approaching completion of all the work in
North Wales, and especially the recent surveys in
Anglesey, where some of the Director-General's
mapping had been revised and modified by his
subordinates, made De la Beche desirous of consulting Ramsay on the ground relative to these
changes.   Accordingly, he asked his lieutentant to
meet him at Holyhead on his return from Ireland.
The diary thus records the meeting :—

'*24th September.*—Got to Holyhead at half-past six,
and found Sir Henry perfectly jolly, but very feeble
on his legs.   We spent an exceedingly pleasant evening together, talking on all sorts of subjects most
unreservedly; I felt quite filial towards him.

'*25th.*—Wet day.   First we had a spread of the
map, with which he was hugely delighted, especially
about the Permian story.   Then I wrote sixteen
letters, and then we had a little walk before dinner.
He looked quite feeble, and like an old man in his
walk.   It quite grieved me to see him, and I felt my
affection growing stronger for him as we walked along,
he leaning heavily on my arm, and using a stout stick
besides.   In the evening we were again very confidential.   He talked about his daughters, their abilities,
Kendall, and all his past life.

' 26*th*.—After breakfast, started in a fly and pair for Amlwch, round about by Cemmaes, etc. He yielded the faults I claimed, and also that the altered rocks were the same as those on the west side of the island. We got to Amlwch by five, and took up our quarters at the "Dinorbin Arms." After dinner he talked of his old friends and acquaintances: Scott, Byron, Madame De Stael, etc. etc., all of whom he knew more or less.

' 27*th*.—Out in a car seeing the gneiss, etc., near the smaller patch of granite. That point I yielded. They are gneiss, and not granite. He was very feeble, and could scarcely, with the help of my arm, crawl along the hillsides, when for a little we put up the car at a farm and walked. But there was a sort of childish good-humour about him that touched me, and I felt as fond of him as I ever did, before he began to get so dodgy with all of us. We spent a most jolly evening together again, he being full of jokes, and making all the servants laugh at his repetitions and kindly talk to them.

' One thing he said to-day amused me much. We were sitting on the sea-beach, eating mutton sandwiches, and watching the action of the waves on the pebbles, when Sir H. said: " I'll tell you what the old gentleman is saying; he's saying: ' Only give me plenty of time,' ha! ha! ha!"

' 28*th*.—Left Amlwch after breakfast in a large car and pair. Beautiful day. Lunched at Pentraith. Sir H. in a sort of happy, amiable, kindly vein all day. We put up at the "George," Menai Bridge. While I am writing he is reading the Bible and commenting on the Flood and other things in what he calls "that funny story." The house being full, we are obliged to take to a double-bedded room that opens directly into the road or yard. Sat latterly in the coffee-room, an English chatterbox, an Indian-looking dragoon, a sensible German, and another man being our fellows. We amused each other pronouncing difficult words for the others to imitate. My Llanfairpwllgwyngyll puzzled all of them.

'29*th*.—Sir H. taken suddenly worse during the night with English cholera, or something like it. He was so bad that by and by he got alarmed, and I jumped out of bed, got a car, drove to Bangor, roused Mr. Charles, the surgeon, expounded the case, and fetched him out with the needful medicines. Miss Roberts in a dreadful way about the bed-room we were in. I brought in Mr. Charles, and Sir H. talked and made him laugh so about what he had eaten and how he felt. . . . The doctor gave him a dose, which almost on the instant put all right.

'I read *Sir Roger de Coverley*, and thought how like the two knights are to each other in many points of character, such as their jollity and harmless humours.

'Between four and six I crossed to Anglesey by the ferry, and saw that old affair of Selwyn's where the Cambrians are supposed to lie unconformably on the older schists. There is every appearance of a fault there, for there is a good 9-inch quartz-lode between them.[1]

'30*th*.—Sir H. quite well and jolly this morning. He vowed that I was his guest here, and that I must not pay any share in the bill, because I would not have stayed had it not been for his illness, so I took an opportunity of slipping into the bar and paying my shot unknown to him.'

[1] See *ante*, p. 172, and *postea*, p. 207.

There were still points of detail and some questions of interpretation of geological structure to be settled in the area mapped by Ramsay and Selwyn in North Wales. Selwyn had gone back to Dolgelli to look into these, and Ramsay joined him there.

'*17th October.*—Held a council with Selwyn on the Shropshire sheets, etc. His work there and here is the perfection of beauty.

'*21st.*—Up in a car as far as the eighth milestone on the Trawsfynydd road; then across the country to Bwlch-drws-Ardudwy. What a magnificent scene! Had a rough climb over Rhinog fach. Let any one who wishes to be convinced of the theory of stratification with subsequent disturbance of beds go there. Their bare and unbroken continuity from top to bottom of the mountains on either side of that savage pass is the grandest sight in Wales.

'*22nd.*—Up to Drws-y-nant by the coach, and then across the hills behind by Dolnallt, Robell fawr, and Benglog. Selwyn made out all his points. How he fights with a bit of ground till he makes it all clear! Truly an admirable workman!

'*3rd November.*—Made some good glacial observations, especially at Capel Curig. Selwyn's semi-scepticism begins to melt.'

These Welsh peregrinations did not pass without including sundry detours to the rectory of Llanfairynghornwy. At last, on the 15th November, on a renewed visit to that remote spot, Ramsay and Miss Louisa Williams were engaged. Among the congratulatory letters which he received regarding this momentous and happy event in his life he carefully preserved that which came to him from his dear old chief. It ran as follows :—

LONDON, 18*th November* 1851.

MY DEAR RAMSAY—Yours of yesterday I have just received in time to say, may you be as happy as I wish you, and may your intended wife value your right sterling honest self as I do. If she does this last, you will be sure of the first. May God prosper you in all ways.—Your ever sincere   H. T. DE LA BECHE.

Ramsay's yearning for a quiet home, with a congenial spirit upon whom he could pour out the full flow of his affectionate nature, was now about to be realised at last. A quotation from his letter to Mrs. Cookman, one of the Dolaucothi family, will best describe how he came to make his choice, and what he himself thought of it :—

'From the first setting of my foot in Wales I was a doomed man. I was fated not to escape from it free. Only think of it! I was done for in the last remaining corner of Wales, where my geological work was to be done, and just about the completion of that work, too. It was in the far north-west corner of Anglesey that I tumbled in head over heels, and was enchained by a maid of Cymru, as thoroughly Welsh as you are, for she speaks, reads, and readily translates Welsh, and, like all Welsh folk, is desperately fond of her country and people. . . . I made their acquaintance accidentally when on a visit to Mr. and Mrs. Fitzgerald of Mapperton (Somersetshire) at Beaumaris. Mrs. Williams asked me to call if I came their way, which I did. I was staying at a bit of a public-house some miles off. Being hospitable folk, they asked me to leave these comfortless quarters and stay with them. I did so ; have been back sundry times since, and behold the result!—a result that considerably surprised both the young lady and myself, but principally the former, for, as is befitting in such cases, I was slain long before she knew it. You ask for a description. Do you suppose I am to be trusted with one? I half suspect not. I'll give you a very little, however, and you can believe as much of it as you like.

'First, then, she is not what you would call pretty, but she is sufficiently pretty to please me. Age about your own, perhaps a trifle younger. She can scarcely be called musical, albeit very fond of it. I mean, she is not much of a player, and but a poor singer, *and she knows it*. But what of these things? I can vouch for her heart and mind. I have met with few girls so well read, and with none so witty. Her love of knowledge is so great, and her memory about ten times as big as mine, that I do consider myself a lucky fellow to have caught a wife that takes an interest in all the pursuits that most interest me, and who did so long before I knew her. But she is not a *blue*. It takes a time to find it out. Then she is so full of mirth and humour, keeping us all laughing. I was always fond of laughing, you know. What more can I say? Her family can't understand how it is possible to live without her, and all the neighbouring poor will miss her almost daily visits.

'The marriage takes place in June, and if the French and Austrians only let the poor Switzers alone, I hope to carry her up the Rhine to Basle, across to Interlaken, thence over by the Jungfrau to the valley of the Rhone, down to Martigny, round Mont Blanc, and down the Arve to Geneva; not galloping, but taking it leisurely, and staying at the pleasantest places for a few days, as the humour seizes us.'

Returning to London towards the end of the year, Ramsay resumed his old place and his old duties.

But everything seemed gilded now by the brightness
that had at last risen upon his domestic prospects.
He opened his course of geological instruction with an
Introductory Lecture ' On the Science of Geology and
its Applications.'   The course began on the 6th January,
and consisted of thirty lectures, given on Tuesday and
Friday.   To make the Museum and its contents more
widely known, and to diffuse a taste for science among
the people, evening lectures to working men were
organised as part of the educational work of the
Jermyn Street establishment.   Each of the six
teachers of the school gave a single evening lecture,
so that the course consisted of six lectures, tickets
being only obtainable by those who could show that
they were truly artisans, and a registration fee of six-
pence being charged for the course.   Afterwards each
teacher gave a course of six lectures.   The instruction
thus afforded, and still continued up to the present
time, has been eminently popular among the class
for which it was designed, large crowds sometimes
assembling in front of the Museum door at the
hour when the tickets for some specially attractive
series of lectures are given out.   In that first winter
of 1851-52 Ramsay chose as his subject ' The Utility
of Geological Maps.'   So much were the lectures
appreciated by the working men that they were
repeated later in the spring.

A few jottings from Ramsay's diary of this period
are here inserted.   Of the meetings of the Geological
Society he writes :—

' 20*th* *February* (1852). — Geological   Society
Anniversary, Willis's Rooms.   President [W. Hop-
kins] pretty well supported — Goulbourne, Sir C.
Lemon, Pusey, Sir H., Lyell, Murchison, etc.   I

observe our body annually creeps higher and higher up the table.   We are now *next* the bigger wigs.

'*25th.*—Good scrimmage between Sedgwick and Murchison on the Lower Silurian and Cambrian question.   It was not an enlivening spectacle.   Sedgwick used very hard words.   Murchison made a spirited and dignified reply.   He appealed to me, and I aided in a speech giving a history of the survey of Wales.

'*24th March.*—Logan's paper [On the Footprints occurring in the Potsdam Sandstone of Canada] and Owen's [Description of the Impressions and Footprints of the *Protichnites* from the Potsdam Sandstone of Canada[1]] passed off well.   Murchison made what Sedgwick called a speech characterised by a sort of bacchanalian joy at the tracks turning out not to be tortoise tracks, and Sedgwick himself rejoiced that the old resting-place of his mind was not disturbed by such a terrible innovation.   He did not like to be too much disturbed.   Lyell was *disappointed*, he said; then Forbes followed, and Owen rebuked them in his reply for entertaining any other feeling than that of joy at an error being corrected, and a scientific truth partly elucidated.   Mantell proposed that they were the tracks of great trilobites, but no one seconded him, or rather every one dissented, Burmeister's paper having gone so far to prove that trilobites had soft membranaceous appendages and no true feet.'

One entry regarding the Royal Institution Friday evenings may be quoted: '*5th March.*—Heard Dr. Mantell give a most amusing lecture on the Iguanodon and other Wealden reptiles.   It was so clever and witty, that throughout it was greeted with rounds of

[1] *Quart. Journ. Geol. Soc.* viii. (1852), pp. 199, 214.

applause.   His raps at Owen through that *Quarterly* article were very characteristic.'

The field-work done by the Local Director this summer included the inspection of the mapping of Worcestershire and adjoining areas.   He in particular traced the boundaries of the Permian breccias between the Bromsgrove Lickey and the Clent Hills, and had his curiosity kindled by the extraordinary character of these rocks.   At intervals he renewed his study of them during the next few years, and came to the conclusion that they proved the existence of Palæozoic glaciers—an announcement which he made at the Liverpool meeting of the British Association in 1854.

The marriage of Professor Ramsay and Miss Louisa Williams took place at Llanfairynghornwy on the 20th July, and two days later he found himself for the first time in his life in a foreign country. Reaching Ostend, the newly-married pair made their way slowly through Belgium to Cologne, up the Rhine to Basle, where they called on Schönbein and supped with Peter Merian, thence to Zurich, and so into the Oberland and the western Alps.   For the first time Ramsay now beheld true mountains and actual glaciers. At the first distant glimpse of the Alps he says that he 'opened his eyes so wide that he feared they never would close again.'   What geologist can ever forget his first transports at such a sight!   How Ramsay's eye caught up the points of special geological interest, while at the same time revelling in all the glories of mountain form, may be shown in a few citations from his diary.

'*7th August.*—As we crossed [the Lake of Lucerne] to Weggis, for the first time we saw a glacier far away towards the summit of the Uri Roth

Stock—I clearly saw the curved transverse crevasses
and a distinct *trainée* of stones. It was an event in
our lives. From Brunnen to Fluelen the contortions
of the rocks exceeded anything I ever saw in the most
intricate old rocks of Wales. Whole mountains were
reversed, 4000 or 5000 feet high. I got a good
notion of these contortions, but very little of the
absolute character of the rocks, for I had no chance of
touching them.

'11*th*.—Were rowed by two men and a woman to
Interlaken. The scenery is so large and grand, the
cliffs so great, the strikes, dips, and contortions of the
great masses of strata so enormous and so grandly
exposed, and the immense slopes of talus below,
scarred with frequent torrents, give such overwhelming
ideas of the incessant effects of atmospheric disintegra-
tion. England, Wales, and Scotland gave me no idea
of it before. At Interlaken we saw descending from
the Breithorn a genuine glacier, not very large apparently, for it was twelve miles off. We had a little
geological scrimmage among the limestones.'

One of the most interesting features of this Swiss
tour was an excursion which Ramsay, leaving his wife
for a couple of days at the Grimsel, made with Dolfuss-
Ausset to the Ober Aar glacier. He gave an account
of this expedition in his article on Swiss and Welsh
glaciers, published seven years later in *Peaks, Passes,
and Glaciers;* but the original narrative in his diary
contains a few personal details which may find a place
here.

'17*th August*.—Went with the guide to find the
" Pavilion " of M. Dolfuss. It was perched upon a rock
some miles off. He is a great gaunt man, and stood on
a rock with a blue bonnet on his head, and a veil

wrapped round it.   As soon as he knew who I was he
hospitably asked me to dine with him, and immediately
after proposed that I should join him in an excursion
to the Ober Aar glacier, which, after a little hesitation,
I acceded to.   So we descended nearly to the lower
end of the Unter Aar glacier, whence I despatched a
note to Louisa, saying I had found an opportunity I
had waited for for thirty-eight years, and that I could
not be back till to-morrow.   We then climbed up by a
brook with four men, and long ere sunset reached the
Ober Aar glacier.   There we had coffee and supper
and buffalo-skins, and by and by my messenger
returned with a delightful note from Louisa.   The
men then cut grass and made a bed in the windowless
hut.   We spread our buffalo-skins upon it, had a glass
of hot brandy and water, put a pipe in our cheeks, and
speedily fell asleep as jolly as sand-boys.

' 18*th.*—Awoke early, long before daylight, a little
damp and sore in the bones.   At half-past three M.
Dolfuss roused himself and blew a blast on his horn,
whereupon all the men got up and lighted two fires,
one in the stove indoors, and the other on a flat stone
outside.   It was a glorious morning ; I thought I had
never seen stars before.   Venus seemed to swim in
the heavens, a ball of light, and not as if a hole had
been punctured in a bluish covering through which
the light shone.   It was glorious, too, to watch the
light gradually growing on the snowy peaks of Ober-
aarhorn and the other peaks that enclosed and nursed
the glacier.   At a quarter to five we started, and were
soon on the ice, five men carrying the burdens.   At
first we were in groups where the ice was solid and
the crevasses distinct.   These required some careful
dodging, though there never was any real danger.

By and by, as we got higher into the regions where
snow had lately fallen, it was needful to be more
cautious. We saw three chamois. We then walked
in a row, following carefully in each other's footsteps,
the foremost man sounding the snow with his pole.
About half-past ten or eleven we reached the snow-
shed where the glacier descends in the other direction
into the valley of Viesch. Then we climbed up on a
flat rock whence Monte Rosa, Mont Cervin, and the
whole of the magnificent panorama of the Alps burst
upon me. The Finsteraarhorn was close at hand,
towering above us in black and white majesty. On
the other side were all the mountains that bound the
valley of the Grimsel, partly hidden by white clouds,
through which the peaks rose as islands. The whole
looked more glorious than I can describe. About one
o'clock we began to descend. On the Grimsel side it
was very rough and steep. [At last from a point 600
or 800 feet above the hotel] M. Dolfuss blew his horn,
and the men gave a yoodle. Met Louisa on the top
of a *roche moutonnée* opposite the inn. Then came
M. Dolfuss, looking tall and rough. We sat together
at dinner, and were exceedingly merry. M. Dolfuss
seemed a great favourite with the landlord and all his
people, and his gaunt yet stately appearance at table
created quite a sensation.'

At Turtmann they were delighted to fall in with
Von Buch and Merian, who were on their way to
Monte Rosa, and would fain have persuaded Ramsay
to accompany them. But he had promised to be back
at his Survey post by a particular date, and so he
reluctantly parted with them, went round by Cha-
mouni, had a scramble on the Mer de Glace, and
by the 2nd September was once more in London.

De la Beche received him with the exclamation, 'Oh, you have come back to the very day; I quite thought you would have taken another week!'

Apart from the general stimulus which a first visit to the Alps gives to a geologist's appreciation of his science, in Ramsay's case a special influence was exerted by the snowfields and glaciers. For the last four years, as we have seen, he had been getting increasingly interested in the various problems presented by the glaciation of Wales. But he had never before actually seen a glacier. The sight of the Swiss glaciers, therefore, quickened his desire to renew the study of the Welsh phenomena, and sent him back with a far more vivid conception of what the conditions must have been in the Ice Age among the hills and valleys of this country. Robert Chambers, to whom, as already remarked, may be assigned a large share in first directing Ramsay's attention to the relics of old glaciers in Britain, received a letter from him soon after he returned from his continental tour, giving some account of what he had seen. In his reply Chambers says: 'I am much gratified in hearing from you at all, and particularly so on account of the late tendency of your studies. In visiting the Alps, and looking at what ice now is doing, you have taken the first step required for the study of ancient glacial action. I could have wished you to take the second (as I consider it) in a trip to Scandinavia. Still, even without that, you may be tolerably prepared for the consideration of the corresponding phenomena in Wales. I have read the abstract of your paper in the *G. Proceedings*,[1] and am really much gratified by the progress you have made in this curious investiga-

---

[1] See *ante*, p. 178.

tion.   Your observations on the drift on the flanks of
Carnedd Llewelyn and Carnedd Dafydd are exceedingly interesting, and indeed the whole article is one
calculated greatly to advance the question.'   There
can be little doubt that this first trip to Switzerland
finally fixed the bent of Ramsay's mind in all his later
geological work.   Though still busy with the many
problems presented by the structure of the older rocks,
these no longer absorbed his attention, nor exercised
that fascination which they had hitherto done.   He
now threw himself more and more into the study of
the origin of the superficial contours of the land, and
among the various agents by which these contours
had been moulded and modified, he specially devoted
himself to the investigation of the work of ice.
Though the bold generalisations of Agassiz in regard
to the former glaciation of Britain had been published
twelve years before, they had met with but small
acceptance among the geologists of Britain.   J. D.
Forbes, Buckland, Darwin, Charles Maclaren, and
Robert Chambers had indeed traced the relics of
vanished glaciers in various mountain groups of Scotland, the Lake District, and Wales.   But a broader
treatment of the subject was needed, and among those
who led the way to this more comprehensive investigation, and who made the Glacial Period one of the
most absorbingly interesting of all the geological ages,
a foremost place must always be assigned to Sir
Andrew Ramsay.

In the course of preparing for the engraver the
various sheets of the map of North Wales, and the Horizontal Sections across the same region, a number of
difficulties presented themselves.   In an area of some
complication, and where the survey had been the

work of several geologists, it was hardly possible that it should be otherwise.   So that portions of the ground required to be revisited, sometimes more than once, and the several surveyors had to meet and discuss the discrepancies or disputed points on the spot.   Much anxious work of this nature occupied the autumn of 1852.   Ramsay took his young wife to Ffestiniog, and from that centre proceeded to clear off all the remaining difficulties up to the Snowdon ground in the north, and Arenig on the east.   Whilst there he was joined by five students from the School of Mines, who came for some initiation into the mysteries of geological surveying.   They included W. T. Blanford, who afterwards rose to distinction in the Geological Survey of India, and is now an active member of the Royal, Geological, Zoological, and Geographical Societies of London ; the late H. F. Blanford, well known for his able contributions to Indian Meteorology ; and H. Bauerman, who afterwards became one of the staff of the Geological Survey of Great Britain.

This Welsh work of completion and revision took longer than had been anticipated.   At the close of 1852 it was not finished.   Selwyn left the Survey at the end of July in that year to take charge of the Geological Survey of Victoria, so that the task of getting the Welsh maps ready for the engraver devolved mainly on Ramsay himself, with the powerful assistance of W. T. Aveline.   The Director-General was waxing more and more impatient.   'MORE (!!!) examinations in North Wales!' he exclaimed to Ramsay ; 'the very sound of such matters sets me adrift. He wished to get rid of Wales, and to have the satisfaction of pushing on the Survey over England.   As nothing delighted him more than to be able to announce a large area of square miles as surveyed in a year, so was he correspondingly chagrined that this prolonged detention of some of the most active members of the staff in Wales should seriously reduce the mileage that could be reported. Much of the summer of 1853 was still required for completing details in some of the Welsh ground, and both Ramsay and Aveline worked hard, sometimes together, but more generally apart. Frequent letters passed between them when they were separate, for Ramsay had set his heart on getting North Wales satisfactorily completed. He had himself been engaged in the work, and felt his credit at stake till he saw the survey finished as fully and accurately as he could achieve. His views were well expressed in a letter to De la Beche, not only regarding North Wales, for which he was himself responsible, but with reference to South Wales, and to all the south-western part of England which had been completed and published before he joined the staff, in a more rapid, less detailed style than had subsequently been gradually introduced, mainly by his own exertions. Writing in the autumn of 1853, when Sir Henry's patience was all but gone, he says (21st November): 'I cannot but think that when, by new lights shining out, omissions or errors are discovered, it is better to mend them, as soon as we know the way, than to leave them open to amateur carpers. It was anything but pleasant the other day to hear of errors and omissions in Malvernia, some of which by accidental visits I knew to be true. You have often spoken of going down to Devon and Cornwall with me to mend the lines there, and I heartily wish the Silurian lines in South Wales and May Hill were mended and brought into harmony with those in the north, by the now easy addition of the dividing line between Upper and Lower Silurian, following out what I did years ago at Builth.

'As these maps stand, their authority is in great part gone, and any one can point out their inconsistencies. I do not, however, even now dream of mending South Wales without special orders, since having been done by others, and before my time, I have no actual responsibility in the matter. When a personal responsibility to you and the public weighs upon me, I cannot rest till I have done my very best as long as I am allowed to do it.'

To his colleague, Salter, he was still more outspoken about the defects of the maps of South Wales: 'When I joined the Survey in 1841, Sir Henry and Phillips did the mapping, and I took lessons and looked on admiringly. I have no doubt that almost all South Wales is bad, Silurian and all. There was no *system* in the work. I suspect my work at St. David's and Fishguard is pretty nearly the best of it. I even separated out the Cambrian, but it was not used. From Builth to Pembroke is a mull, Llandovery and all. Certain I am that Sir Henry had no ground for putting my Llandeilos above the Castell Crag Gwyddon rock. I had nothing to do with it. Sir Henry began to map it, and left it off unfinished. The whole is only about ten stages better than Devon and Cornwall.'

Some of the maps and sheets of Horizontal Sections of North Wales having now been published, Ramsay took a useful step in the spring of 1853 by reading to the Geological Society a brief outline of the general succession of rocks and geological structure of the region, so far as these had been determined by the Survey. In this paper he passed over the contention of his colleague Selwyn, which, as we have seen (pp. 172, 192), was also his own original impression, that the Cambrian rocks of Anglesey are underlain by a far more ancient series of schists. He now published his belief that these schists are the metamorphosed equivalents of the Barmouth and Harlech grits, and the Llanberis and Penrhyn sandstones and slates—an opinion which he maintained to the last. The paper is interesting as the earliest account of the successive groups in the older Palæozoic rocks of North Wales, as finally worked out by the Geological Survey.

Among the incidents of the summer of 1853 the most notable in Ramsay's life was the birth of a daughter on the 3rd June at Beaumaris. It gratified him to think that in Wales, which had almost become his adopted country, and which, by the ties of marriage, had now grown doubly dear to him, his child should have been born.

An important departmental change this year affected the Geological Survey. Once again it was transferred to a new set of masters. The exchange arose in this wise. One of the consequences of the Great Exhibition was the impulse given to the recognition of the importance of Science in national progress. In 1853 a comprehensive scheme was carried out by Lord Aberdeen's Government, whereby a 'Department of Science and Art' was established under the charge of the Board of Trade, of which the President at that time was Mr. Cardwell. The control of the Geological Survey, the Museum of Practical Geology, and the School of Mines was transferred from the office of Woods to this new department. Three years later, that is in 1856, another change was made, whereby the Department of Science and Art was transferred to the Privy Council, and was administered by the Lord President of the Council, assisted by a member of the Privy Council, who is called the Vice-President of the Committee of Council on Education.   This arrangement is still maintained.

CHAPTER VII

THE GEOLOGICAL SURVEY IN SCOTLAND

THE general awakening of the country, after the Great Exhibition of 1851, to the national importance of cultivating science for industrial if not for theoretical purposes, showed itself in Scotland, among other ways, in an agitation for the extension of the Geological Survey to that part of the United Kingdom. The movement never needed to be vigorously pushed, for the mapping of Scotland had from the beginning been contemplated as part of the duty of the Geological Survey of Great Britain. But two serious impediments had hitherto stood in the way of even making a commencement of the work. In the first place, the Ordnance Survey of Scotland was so far behind that the maps, on which alone a geological investigation could be properly based, had not been available ; and in the second place, even if maps could have been obtained, the staff of the Geological Survey was so small that it was hardly possible to spare any officers for breaking ground in Scotland. In response to the agitation on the subject, De la Beche instructed Ramsay to go to Scotland and see for himself the actual state of affairs. Writing to his lieutenant on the 26th July 1853, he said : ' The Survey (Geological) of Scotland has been long ordered (*vide* votes in Parliament), but we

P

could never enter upon it, because there were no proper maps, not enough to continue the work upon if we had commenced it. *Now* let us see if there are the needful maps, and a fair prospect of not being compelled to break the work off if commenced. You have been sent in your official capacity to examine and report to me on this matter, so that it may be seen how far it may be expedient to commence the Survey of Scotland next year, your report and other needful inquiries enabling those who will have to decide on the subject to obtain a correct view of it.'

In accordance with these instructions, Ramsay went to Scotland in that year, and spent the month of August there making the necessary inquiries, and at the same time visiting some of his relatives and old friends. He found that in the central part of Scotland, where upon the coal-bearing formations and among the great industrial districts it was desirable to begin, the Ordnance maps were exceedingly backward, but that there was a prospect of obtaining unfinished proofs of them in the course of next year. It was subsequently arranged that he himself should go north and begin the geological survey in the summer or autumn of 1854. His letters during this brief expedition north of the Tweed show how impossible it was to escape from the pressure of official work. Sir Henry was getting old and less able than he had been to keep himself in touch with the field-work, though to the last he continued his tours of inspection, and even in the summer of 1854 was down in County Cork exploring with Jukes the coast-line about Bearhaven. But it was essential for the progress of the service that Ramsay should constantly keep himself in communication with his men. Thus he dragged at each remove a lengthening chain of correspondence. A few extracts from his notes to Aveline, some of them written when he was really on holiday visits in Scotland, may here be given. 'What has become of you? What are the prospects of the work? [Completion of part of the Welsh ground.] Is it done or nearly done, or does it look as if it would be done; and have you been able to solve your difficulties? Sir H. wanted to disturb you. I wrote trying to stave him off. . . . I have been away a day and night among the islands of the Forth in a steamer belonging to the Commissioners of Northern Lights, and landed on the Bell Rock. It is Old Red Sandstone and twelve miles from shore. I will send your sections, maps, etc., in a day or two. It is not easy to find quiet here. When I get to Hamilton I will send you a Cader sheet; I have none here. Yesterday I got some fine specimens of foliated mica and chlorite schists by Loch Lomond and Arrochar. The *glacial* phenomena beat anything I ever saw. It is wonderful.'

At this point of the narrative, when the operations of the Geological Survey are to be described in Scotland, it may be of advantage to look for a moment at the state of the progress of the work at that time in England. The whole of Wales had been completed and published, together with Cornwall, Devon, Somerset, Dorset, Wiltshire, Worcestershire, Herefordshire, Warwickshire, Staffordshire, Shropshire, and Derbyshire. Portions of some other counties had also been published, and the field-work was now being pushed into Lancashire and Yorkshire, north of a line drawn from Liverpool to Sheffield, and into the counties of Nottingham, Leicester, Northampton, Oxford, Buckingham, Berkshire, Hampshire, Surrey, Sussex, and Kent.

The staff at this time in England consisted of Sir Henry de la Beche, Director-General; A. C. Ramsay, Local Director for Great Britain; W. T. Aveline and H. W. Bristow, geologists; W. H. Baily, J. W. Salter, H. H. Howell, and E. Hull, assistant geologists; R. Gibbs, general assistant; and J. Rhind. The Irish staff, under the Local Director, J. Beete Jukes, consisted of W. L. Willson, geologist; A. Wyley, G. V. Dunoyer, F. J. Foot, G. H. Kinahan, S. Medlicott, and J. O'Kelly, assistant geologists; and James Flanagan and Pierce Hoskins, specimen-collectors.

The maps and sections of North Wales having been completed and published, the next labour was the preparation of a connected description of the geology of that region. No one but Ramsay could undertake this onerous duty. He was familiar with the Principality from Holyhead to Caermarthen; many square miles of it he had himself surveyed, and he had inspected at various times almost all the rest of the ground. Still he could not be expected to know all the details of tracts which he had not personally mapped. He accordingly applied to his colleagues, more especially to Aveline, Selwyn, Jukes, and Salter, for notes regarding their several districts, and these, together with the memoranda he had himself made, he proceeded to weave into a continuous Memoir. This task continued to be his chief indoor labour for some years. Ill-health eventually seriously delayed its completion, and the Memoir did not finally make its appearance until some twelve years later. To do the editorial work of this volume satisfactorily it was desirable that he should plant himself for a time in some central part of the region to be described, so that he might easily verify any doubtful points by actually visiting the ground. Accordingly he took up his quarters during part of the summer of 1854 at his old station, Llanberis. A few passages from his letters will give a picture of his life and work there. Writing to Aveline on the 7th June he tells how an illness there had retarded him, and adds : ' If I can walk a mile or two to-day I shall try several more to-morrow, and if that succeed, leave this suddenly. I have now got to that part of the Memoir that deals with the Bala beds and Caradoc, from Mallwyd all round by Yspytty Evan to Conway, and am especially hard up for information in places. Will you turn up your note-books and copy out any descriptions of the structure of the rocks, etc. ? Never mind digesting them into regular description more than you like ; only give me any notes you have, and I'll quote them in your own words when I can. I am at present incorporating what Jukes gave me about the Bala Limestone and ash, and of course what I chiefly want is anything about the rocks above that.' Again on the 16th he writes to the same correspondent : ' On Monday, after an early dinner, I took a gig to the top of the Pass, and then started across the hills for Ffestiniog, in part over a bit of country I had not been on before. I was anxious to see it before describing its rocks. I passed by the lakes called Llyniau, under the west end of Moel Siabod, and through the upper part of Dolwydd-elan, and down by Manod-bach. The greenstones are right. I reached Martha's about ten at night, and got a hearty reception and supper. Next morning a commercial gent gave me a lift to Trawsfynydd, where I struck into the country, and went over Craig-y-Das-Eithen and down by the back of Penmaen and Cwm Eisen to Tyn-y-groes, where I dined, and went on to

Dolgelli. Next morning Byers and I had a scramble till 6 P.M., when I got a return car as far as the cross-roads near Ffestiniog. I slept at Ffestiniog, and next morning walked up Cwm Orthin and over Cynicht to Cwm Gwynant, near Beddgelert. I had never been on Cynicht before, and learned a few things for the Memoir. I then walked to the top of the Pass, where I found a return car and reached Llanberis at half-past eight. It was a day's work that.'

To Salter, on the 28th of the same month, he writes : ' Though deep in ice ' (he had been preparing a paper on the Welsh glaciers for the British Association), 'that was only occasional work ; I am deeper in the Memoir, and have got a great deal done. In consequence of last year's finishing strokes to the traps, between the Bala road and Arenig bach, I was obliged to re-write a good part and to re-arrange the order of description. That is all done now, and much more, including, of course, the Lingula Flags, which, however, I do not consider finished without your advice thereanent again. Your notes are of the highest value. I have used them freely, quoting you, but putting them necessarily into more current or running English. This will be submitted to you in good time. I have also used Jukes's notes about the Bala country, altering and abolishing a little, and adding considerably. I think I must send you that to read and comment upon soon, for I repeat that you can give me much help both there and in the Berwyns. Not a soul has given me a single note about the latter.'

To the same correspondent, on the 13th August, he sends thanks for congratulations on the birth of a son on the 6th of that month, and adds : ' I hope to be at Llanfairynghornwy for a day by Thursday at latest *en route* to Ireland, where I want to have a touch at the six-inch maps before beginning in Scotland. . . . This Memoir I do not mean to say will be done (with so much at present on my shoulders), but I do hope to finish it next winter and early spring. . . . Were we to go down together next spring [into Wales], and possibly take Jukes and Aveline with us, then putting all our experience and knowledge together, we might through the year produce such results as would throw a strong and steady light on Ireland. I believe that no man single-handed could do so in two or three months, and I believe that if you ask Forbes (if he be in London) he will agree with me, for our opinions, I have observed, always pretty well coincided in such matters.'

The journey to Ireland referred to in the foregoing extract was chiefly for the purpose of personally inspecting the system of mapping followed by Jukes and his colleagues. From the commencement of the Survey in Ireland maps on the scale of six inches to a mile had been used as the field-maps. In England and Wales the general map on the scale of one inch to a mile had alone been available for geological purposes ; but as the six-inch scale was now extended to Scotland, it was proposed to carry on the geological investigation of that part of the United Kingdom on the larger scale. It will be readily understood that the substitution of a map embracing thirty-six times the area of that which had been used in England and Wales would necessitate numerous modifications of the system of mapping applicable to the smaller scale ; much more detail could be expressed and abundant notes could be inserted, which in the case of the one-inch scale required to be written in the note-book.

De la Beche, though he had consented to the adoption of the larger scale in Scotland, was rather inclined to disparage it. But as that scale was to be employed, Ramsay clearly realised that it was his duty to profit by the experience of those of his colleagues who had been using it for years. 'It would be a great mistake on my part,' he wrote to his chief, 'to omit seeing what they do, and how they do it, in Ireland. It does not follow that the same rules should be applied in Scotland; but whether or no, I want to see how they keep, cut, use, and abuse their maps, what their portfolios are like, how they handle them in the field, and twenty other things that may save us much time and trouble in Scotland, and which only eyesight can instruct upon.'

De la Beche's bodily and mental powers were visibly failing, though his natural gaiety of temperament showed little abatement. His declining vigour appeared more especially in the uncertainty and vacillation of his official decisions. He had both verbally and in writing agreed that Ramsay should begin the survey of Scotland, but afterwards, when all the arrangements had been made, he was afraid to go on with the proposal, lest there might be some questioning on the subject at headquarters. Ramsay, however, knowing how fully the matter had been discussed and approved by Sir Henry, determined to persevere in the course which had been fixed upon. In pursuance of that resolution he crossed to Ireland to see Jukes and his men at work, and at the same time to have one more conference with the chief, who had joined the survey party in the south-west of County Cork. The story of the interview is told in a letter to Mrs. Ramsay of the 25th August: 'Yesterday after breakfast, Jukes,

JOSEPH BEETE JUKES

Willson, Kinahan, and I drove in a car to Glengariff; Mrs. Jukes rode out with us for a mile or two on Dolly, and two dogs were also of the party, Carlo the setter, and Tommy the Scotch terrier. We dropped the other men *en route*, and Jukes and I drove on to Glengariff. It is not a town, but a tourists' inn in a lovely valley. It puts one in mind of Loch Lomond, only the water is salt and the hills not so high. There we found Sir Henry, Rose, Kendall, and Carry Smyth. After shaking hands, "So you've come here," quoth the Governor. "Yes, I could not help it!" "I think you might," and then he showed me how it was impossible to begin Scotland; he had no objections to my going down to open the ground (not to *map* it), but it was impossible to authorise any one accompanying me, for Cardwell had said this and that and the other thing. I asked him the meaning of his letters urging me to go down and get something published. Just at that moment a question arose about Kennedy and Medlicott. Jukes and Sir H. had a long discussion, during which I had ample time to quiet all vestiges of rising wrath, and to arrange my plan of argument, which was so effectively done, that when Sir H. and I set-to again, I got him to agree to everything I wanted. I go down when I please, and get Aveline to follow me! So far well, with three cheers for diplomacy and honesty combined! Sir H. is more to be pitied than railed against; for his mind is far, far gone, though you would not think so under ordinary circumstances.'

The few days spent with Jukes and his colleagues gave Ramsay a good idea of the way in which geological mapping on the six-inch scale could be carried on, while at the same time it presented him for the first time in his life with some vivid examples of Irish scenery and Irish manners and habits. Writing to his wife from near Bantry on the 27th August he says : 'The weather still continues splendid. We had a long walking and car-ing day yesterday through a beautiful country, wild and rocky. They call it here cultivated in places; but to my eye a great part of it is a sad spectacle. You see as many houses without as with roofs, and few gates swing on their hinges. But the people are fine-looking, those of them that get plenty to eat, tall and stout, with long arms and upright gait. The women are often pretty, and they can do what few English women can, they walk erect and graceful, with long steps. They do not hobble or amble or mince ; they walk.

'I am just about starting for Glengariff, and to-morrow will be at Killarney. I bathe every morning, and am quite recovering all my swimming powers. I swim right away out to sea on my face, and return on my back by way of a change.'

After coming back from Ireland he wrote to Aveline on the 1st September : ' I purpose starting for Scotland next week, and think of beginning about Dunbar, but am not as yet certain. I shall return to the meeting of the British Association at Liverpool, and some time after that (not very long) I certainly expect to want you in Scotland, both that we may make a good beginning, and also that when I am obliged to leave (going back and forward) I may have a representative at work.'

The British Association met this year at Liverpool, and Edward Forbes, who had recently left the staff of the Geological Survey to succeed his old master, Jameson, in the Natural History Chair in the

University of Edinburgh, was President of the Geological section. Ramsay came to the meeting to support his friend and read two papers. In one of these he made known the startling conclusion to which he had come, that in Permian time glaciers existed in this country, and had left behind them the remarkable breccias and boulder-beds of the Malvern and Abberley hills. The progress of the Survey across Worcester-shire into the central parts of the midlands had given him ample opportunity of studying these breccias. He still further elaborated his observations, and com-municated them in the following February (1855) to the Geological Society. He was now in the full tide of glacial enthusiasm. The old Welsh glaciers had acquired renewed interest from his experiences in Switzerland, and he had endeavoured to track their course and measure their thickness by the markings they had left upon the rocks. He obtained renewed proofs of the two periods of glaciation he had already indicated, and now found that at the time of its greatest extension the ice had actually passed across some of the larger valleys, such as those of Llanberis and Nant Francon. He ascertained by direct measure-ments of the heights of the striation on the rocks that the ice of the greater glaciers was about 1300 feet thick.

After much delay the Geological Survey was at last extended to Scotland in the autumn of 1854, and the Local Director himself undertook the task of beginning the work. The state of the Ordnance Survey county maps on the six-inch scale left little choice as to the district where geological work should be started. Ramsay finally determined to commence on the coast at Dunbar, where he could trace in the base of the Carboniferous system, and, working gradually westwards, might clear the way for the further prosecution of the work by his staff next year. His rambles led him into some of the lonely valleys of the Lammermuir Hills and along the picturesque shores of East Lothian. He had had no quiet geological work in Scotland since his old Arran days, and there was so much of interest and novelty in thus breaking ground for the Geological Survey of Scotland that, repressing his strong desire to be back once more with his young wife and children, he persevered with the work until the keen frosts of December drove him at last southward. A few pictures of this working-season at Dunbar may be gleaned from his letters to Mrs. Ramsay.

'The Old Red Conglomerate here [among the Lammermuir Hills] is the most wonderful deposit I ever saw, and horribly icy-looking. It is so soft, too, you might dig it out with a pickaxe. The greater part of it is almost indistinguishable from Drift. Examine the enclosed stones and give me an account of their colours. Merely write me their colours and then throw them away. Whatever colours they are, it does not in any way affect my mapping, but it would be a satisfaction to be *certified* on the subject. Are they grey mostly, and is there any purple and green? [1] They are fragments from the stones that make up this tremendous brecciated conglomerate on which I walked on Friday all day without any prospect of getting to the end of it. It forms great wild, heathy hills, stretching far away south into Berwickshire. It is so

---

[1] In explanation of these directions it requires to be stated that Ramsay was to a considerable extent 'colour-blind.' He was often made conscious of this defect when discussing the Survey maps with his colleagues, for he could not distinguish between some of the colours, reds and greens being especially mistaken.

incoherent that it is everywhere traversed by the most remarkable ravines, deep, steep, and often without water in them. These have been made by successions of winter floods, and sometimes in the course of ages, the drainage having taken a new direction, they have become permanently dry.'

'Surely this place is "Cranford." I by no means understand the constitution of its society. Yesterday the only person besides myself at dinner with the —— was a Mr. Combes. He was a stout little gray-haired man in black, who from his appearance might be a clergyman with a black neckerchief, a schoolmaster, a professor in a Scotch college, a physician, a surgeon, a country gentleman, a retired merchant, a first-class skipper, or anything you like, not great or noble. Well, the conversation got animated, and our host made an occasional *mal à propos* remark and thoroughly enjoyed the talking. The little man discussed history, English, Scotch, and Roman, the styles and merits of Hume, Smollet, Robertson, Gibbon, and Scott, of the *Pictorial History*, of Mackintosh, Fox, and Macaulay, of the novel-writers, including Fielding, Smollet, Miss Burney, Misses Porter, and all the moderns, the history of poetry as shown in the writings of Dryden, Pope, Burns, Ramsay, Tannahill, Fergusson, etc. And, besides, he had been in India, and had voyaged about, that was clear. Well, he and I walked home ; we shook hands, and he turned into a house in the street, and I looked above the door, and saw thereon COMBES, CANDLEMAKER !'[1]

'With a large section of society intellect is not to

---

[1] This story ought to end here; but Ramsay afterwards found out that his enigmatical companion had been a surgeon in the old navy of the East India Company.

be endured, especially if it dare in any way to think a little differently from the common herd. Do you know, it costs me no small trouble to keep out of hot water even with —— ? Our style of thought is so utterly diverse that there is almost no chance of our agreeing on any possible subject. I never heard an interesting conversation in their house except from others. But I have a great respect for them and much affection. How curious is the difference when one gets away among men of learning and science! They do not see merely the outside. They reflect and reason, and whether correctly or not, still it *is* reflection and reason. I never saw Goodsir[1] but once before, and that for five minutes. We were at once friends, and I feel that I love him, for we have a community of thought, though our sciences are quite different. He is so unassuming, simple in manner, gentle, and kind. I have much to learn from some of these men.'

'This is a bright sunny day, with a westerly wind and white waves dashing on the red cliffs and islets below the Castle. The Castle is the merest fragment of a ruin—a few walls some three or four yards thick on a rocky promontory. Lauderdale House looks upon them, still entire, but deserted by the Earl, the windows closed up or broken. It faces the sea, and its back looks down the long street very drearily. The family left the town when the Reform Bill put an end to their borough influence. A winged sphinx sits on the roof, and wonders how long it will be before it will fall in.'

[1] John Goodsir, born 1814, died 1867 ; one of the greatest anatomists of his day, was an intimate friend of Edward Forbes. He was Professor of Anatomy in the University of Edinburgh.

'The wind is rising strongly from the east. The sea gets white. The square-sailed fishing-boats come scudding in for shelter, and two large three-masted steam-screws are scudding up the Firth with all sails set forward. I write during breakfast. One of the boats seems unable to make Dunbar Harbour, and is running for shelter into Belhaven Bay. The men's wives are looking out across the walls. . . . That little boat has beat up to windward after all. She is now under shelter of the Castle; down goes the square sail, off go the oilskins, and out go four oars, and she will be in dock in a twinkling.

'Heigh ho! A boat after all has been upset on Tyne Sands. Three men are drowned and two saved. I saw a woman pass crying, and afterwards a sailor, looking very grave. I feared something.

'The tumult of the waves is wonderful to look at. They come rolling in, and swallow up the rocky islets that guard the shore, breaking over them in great sheets of white water. The roar, the great masses of spray, and the labouring vessels dimly seen scudding up the Firth—everything seems to bespeak disaster.'

Before he left Dunbar the Director had completed about a third of the area he had assigned to himself to be mapped from that station. The Geological Survey of Scotland was thus fairly launched. Ramsay, however, was never again able to find time to resume the mapping of any area north of the Tweed.[1] All that his increasing official duties permitted him to accomplish was to come down year after year and inspect the work of his colleagues, completed and in progress.

---

[1] The only time that he took the maps himself into the field was some ten years later, when, while spending a few days with his friend Mr. J. Carrick Moore in Wigtownshire, he mapped the end of the peninsula which terminates in Corsewall Point.

While stationed at Dunbar a calamity befell him which all through the rest of his life he never ceased to deplore. His cherished friend, Edward Forbes, who only a few short months before had succeeded to the Professorship of Natural History at Edinburgh, and who had presided so genially and actively over the Geological Section at the recent meeting of the British Association at Liverpool, was cut off on the 18th November after only a few days' illness. Ramsay thus announces the sad news to his wife : ' I have just received a brief note from James Forbes telling me that Edward died on Saturday at a quarter to five. I can scarce realise it. My grief breaks out in short fits, and then I struggle to suppress its signs. O Louisa! what a terrible blow, and how seemingly inscrutable. In the flower of life, with a dear wife and children, and in the new opening of another phase of a great and useful career! The more I think of him, the more I feel that, next to you, he has exercised more influence on me than any other person I ever knew. He was so earnest and so good. I wish I may be able in some small degree to imitate his worth. We had much in common, but, as a man of science, his station was much greater than mine can ever be. Forbes never lost a friend. His goodness as well as his greatness make him so universally lamented.'

On returning to London for the winter of 1854-55, the various members of the Geological Survey were concerned to find how greatly their esteemed chief had altered for the worse since they had last seen him. Ramsay notes : ' Sir Henry is wofully changed. He is so feeble now that he has to be carried in in his chair, and wheeled to his room. He looks shrunk, and his face is doubly lined ; neither is its expression

EDWARD FORBES

improved.' Nevertheless, the veteran stuck to his post, presided over meetings of the teachers of the school, had all the latest Survey maps submitted to him, and took the keenest interest still in all the field-work. He was specially pleased with the field-maps on which Ramsay had traced the first beginnings of the survey of Scotland, and expressed his approval not only to him, but also to his colleagues. 'Ramsay is advancing,' he said, 'and showing much official aptitude.' Not only so, but he manifested much interest in the Local Director's glacial work, of which he had previously been inclined to make light. 'He was delighted,' says Ramsay, 'with the Swiss and Welsh moraine matter compared side by side, and actually gave some obscure hints, as if I should be obliged some day to be over a good deal in Ireland, apparently meaning that I should by and by have the charge of both Surveys.'

De la Beche had often previously spoken to Ramsay on the subject of the next Director-General. In their more confidential moments he had assured him that his 'geological son' should succeed him, and that he had put that wish so definitely in writing that he felt certain that the Government would follow his advice. For years, therefore, Ramsay had come to regard the reversion of the office as secured to himself. But during the year that preceded the time at which we have now arrived various circumstances had occurred to shake the confidence of his belief on this matter. Sir Henry's failing powers, mental as well as bodily, had led him to take, on more than one occasion, a course which Ramsay, feeling strongly that it would be detrimental to the best interests of the service, opposed as firmly, though of course as

Q

courteously, as he could.   The chief was now apt
to be impatient and somewhat exacting, as well as
inclined perhaps to push official routine and regulation
further than his able subordinates thought necessary or
desirable.   Ramsay had an exceedingly difficult and
delicate task to discharge.   He sympathised with his
colleagues, and was entirely loyal to the Survey, at the
same time that he had a strong affection for his chief,
and a keen sense of the duty of subordination and
discipline.   So long as he could be side by side with
Sir Henry, there was little risk of serious difference.
The veteran's sense of what was just and fitting was
so strong, and his confidence in his lieutenant so
entire, that he soon came to an amicable agreement
when they argued a question together.   But Ramsay
had occasion to be much in the field, and his place
was apt to be taken by other counsellors, whose
advice did not always coincide with his.   Certain it is
that towards the end of 1854 De la Beche, feeling
that his own days were numbered, and being desirous
of playing some part in the nomination of a successor,
perhaps also somewhat displeased because Ramsay
had recently withstood the promulgation of a vexatious ordinance, fixed his thoughts for a time upon a
geologist other than his lieutenant as the proper
person to succeed him, and there is reason to believe
that he privately communicated his views on this
matter to the Minister in whose hands the appointment lay.

Up to the very end Sir Henry came to the
Museum, even though he could not leave the chair in
which he was wheeled into the building, and his loud
voice and hearty laugh could be heard all over the
place.   He had still his joke for each member of the

staff, and his kindly word of inquiry and encouragement for the attendants and cleaners. 'Well, Mr.
——, are you happy?' he would ask of some new-
comer, as he was wheeled across the Museum floor to
his own room facing Piccadilly. He appeared for the
last time on Wednesday, the 11th April. It was
his intention to be back on the following Saturday,
but he became rapidly worse next day, and died on
the morning of Friday, the 13th.

From the allusions which have been made in the
foregoing chapters some more or less adequate picture
may be formed of the character and work of this
remarkable man. His scientific achievements placed
him in the very front rank of English geologists. His
kindliness of heart and gleefulness of spirit endeared
him to all who came into close contact with him. The
very failings which have been already indicated did
not alienate the affectionate regard of his associates.
Even Ramsay, who perhaps suffered more than any
one else from these failings, loved him to the last, and
mourned for him as for one of the most leal-hearted
friends he had ever had.

It is not necessary or desirable for the purposes of
this biography to enter into the details of the appointment of a successor to De la Beche in the Geological
Survey and the establishment at Jermyn Street. It is
enough to say that Ramsay soon saw that the hopes
he had cherished for so many years were doomed to
utter disappointment. He found, moreover, that
vigorous efforts were being made in favour of a most
estimable man of good family, but possessing only a
very slender acquaintance with geology. As there
seemed some possibility that these efforts might be
successful, and that the Survey, Museum, and School

might thus be placed in most incompetent hands, Ramsay proposed at a meeting of the professors that Sir Roderick Murchison should be their next chief. This suggestion being agreed to, was communicated to Mr. Cardwell, who approved of it. Thereafter, and apparently with a view of strengthening the application in favour of Murchison's appointment, a memorial urging his claims was prepared by Dr. Fitton, who obtained numerous signatures to it from leading men of science, including Sedgwick, and this document was sent in to Government. Within less than a month from De la Beche's death Murchison was appointed to succeed to the office, and he entered on his duties on the 5th May. Ramsay did not allow even his intimate friends to know how bitter was his disappointment at the loss of the prize which his own chief had taught him to believe would certainly be his. It was one of the most trying episodes in his life. But to the distinguished geologist who now, as it were, supplanted him he brought the most unswerving loyalty, and remained his faithful colleague up to the last.

Before this important matter was definitely settled Ramsay had started once more for the field. There were two departments of the surveying now in progress in which he took a special interest—the Permian mapping of the midlands and the revision of parts of Wales. He first went over the recent Permian work, and while so doing sought for further light upon the question of Permian glaciers, which he had brought fully before the Geological Society on the 21st February in this year. The subject had been rather laughed at by De la Beche, who said, 'As to the scratching of breccia fragments—"'tis their nature to"

—a tumble-down house will give plenty of them; and then as to old localities for the fragments, independently of not having cakes which have been eaten, who the dickens, in such places, can say what rocks are beneath the sprawl of New Reds?' Lyell, however, took a much more serious view of the matter, and with that eager enthusiasm so characteristic of him, threw himself into it, and endeavoured to master all the evidence. He asked Ramsay to go óver the ground with him, and the request was readily granted. 'Lyell is brimful of these Permian glaciers,' Ramsay wrote to Aveline; and after taking the author of the *Principles of Geology* over the ground, he tells in his diary how at dinner Lyell, who had been ruminating on the subject for some days, at last declared that to his mind and eye the breccias told of river-ice rather than of glaciers and icebergs. This dinner took place on the eve of another continental journey, which Ramsay had planned for the purpose of comparing some of the more notable breccias of Germany with those which he had been studying in England. Starting with Mrs. Ramsay at the beginning of August, they made their way by Heidelberg to Eisenach, spent some ten days there examining the Rothliegendes and other formations of Thuringia, and made a brief visit to Berlin.

The revision of some of the Survey work in Wales was a subject that had lain very near to Ramsay's heart for some years before the time at which this narrative has arrived. We have seen how imperfect he knew the maps of South Wales to be, owing to the want of any proper subdivisions among the older Palæozoic formations. At the time when that ground was mapped the importance of such subdivisions had

not been recognised.   It was enough, in Sir Henry De
la Beche's opinion, if these ancient rocks were distinguished on the maps by some one common colour.
But as the work advanced northwards, and the true
significance of the labours of Murchison and Sedgwick
began to be perceived, it was seen to be eminently
desirable to separate at least some of the larger groups.
The great break between Lower and Upper Silurian,
which Ramsay had detected near Builth, was one of
which he early saw the importance.   Sedgwick and
M'Coy had shown in 1852 that rocks which had been
grouped by the Survey with the Caradoc sandstone in
the Lower Silurian series contained such an assemblage
of fossils as linked them rather with the Upper Silurian.
Hence it was that Ramsay, who felt himself responsible
for the mapping of Shropshire and the adjacent tracts
of Wales, and was anxious that the Survey maps
should be made as accurate as possible, deputed W.
T. Aveline and J. W. Salter to re-examine that region.
The result of the labours of these two members of the
staff was to establish beyond any doubt that the break
between the Lower and Upper Silurian series of
formations in that part of Britain was complete, and
that the so-called 'Caradoc' of Murchison and 'Bala'
of Sedgwick were palæontological equivalents, the one
of the other.   It was then evident that the boundary-
lines thus established, and which were put on the
Survey maps, would need to be carried into South
Wales, where hitherto no attempt had been made to
show any stratigraphical subdivisions in the series of
formations below the Wenlock group.   And this
southward extension became all the more necessary
after Aveline had separated out the 'Tarannon shales'
below the Wenlock group, and had shown what a

persistent zone they formed in North Wales. Hence
at last, and after much objection on De la Beche's
part, who, as we have seen, was weary of these repeated re-examinations of Wales, Ramsay obtained his
chief's authority in 1855 to send Aveline into South
Wales for the purpose of inserting the more glaring
omissions on the maps and improving the representation of the associated igneous rocks. How keen was
the interest that Ramsay took in this work is shown
by the voluminous correspondence which he carried
on with his colleague in Wales. For weeks together
every second or third day he would write to Aveline
when at work in North Wales, commenting on the
last report received from him, and suggesting localities
for re-examination or points to be kept in view. Two
of these letters may be cited as examples.

LONDON, 12*th April* 1855.

MY DEAR TALBOT—I have Sir Henry's consent for
you to have a turn in Pembrokeshire when you have
done with the Wenlock line. The business will be
first to do the Cambrian line, and secondly to walk
across each bit of trap in the country, tap it and
note down whether it is hornblendic or felspathic,
melted or felspathic ash. Except errors stare you in
the face, don't bother about them ; but of that more.
This job is urgent now, because Sir Henry has decided
to change the colouring, and to colour all greenstones
green, and all felspathic traps and granites shades of
red. I am glad of it, for it gives us a good opportunity of improving the Pembrokeshire maps, and with
Sir H.'s determination about the colouring, there is no
escape from the necessity of looking at it. I suspect,
indeed I know, there will be mapping to do in the

great St. David's lump of trap. There are all kinds there, unseparated and only indicated by one colour. Once you get to Llandovery I think you will rapidly finish the Wenlock line. But don't neglect to map such sandstones as are needful.—Ever sincerely,

ANDW. C. RAMSAY.[1]

ASHBY DE LA ZOUCHE, 17th June 1855.

MY DEAR TALBOOTS—The Pembrokeshire fragments [of map] will follow this by to-morrow's post. The plain green [colour on the map] is intended to represent greenstone, and the striped green, felspathic trap. In the parcel are fragments drawn by me on an enlarged scale for the purpose of mapping when on the ground. *I know that all the traps between St. David's Head and Pen-berry are intrusive greenstones,* and you need not visit them unless you like. I fear that the trap on which St. David's stands will want looking to and separation into kinds all the way from east to west. I think you will find most of the long strips greenstone, but I recollect that some of them between Aber - pwll (four miles north - east of St. David's Head) and Mathry are felspathic ash. I once stayed a week at a public-house at Mathry.

St. David's Island is mostly greenstones, but some felspar; Lower Solva and Whitchurch, felspar and quartz, I think; in fact, a granite without its mica. I think Trefgarn is greenstone, but am not sure. I suspect there are both kinds near St. Dogwells. At Wolfe's Castle there was a public-house, where I put

---

[1] A melancholy interest attaches to this letter, for it intimates the last official act of De la Beche's life. He had been at the Museum the day before, and given the consent above referred to; but before the letter could reach its destination 'our dear old governor, whose like we shall never see again,' as Ramsay wrote, had passed away.

up a horse daily. Sir Henry did most of the Preseley traps. You will find remarks of his on the maps that may help you a little.

Are the Cambrians coloured on the copy of the Map you have with you? On Nos. 1 to 5 you will find some useful hints about them near St. David's, and on Nos. 6 and 7 you will find them mapped in some sort of way pretty carefully. You must judge for yourself whether it will now and then be needful to cut it fine and alter old lines. I have a notion that most of mine are good, viz. all those west of St. Lawrence and Mathry; all between Little Newcastle and Trefgarn; also those between Fishguard, Newport, and the River Gwaen. The rest I know little about, though I did some of them.

Finally, do not spare horse-flesh or car-hire to do it quickly.

As soon as it was possible for him to join his trusted colleague, Ramsay made his way into Pembrokeshire. He had not been back there since his early days in the Survey. The inspection had thus a double interest for him. To Mrs. Ramsay from Dale, on the 13th November, he writes : ' We were to-day at West Angle Bay in a sailing boat, which I had the satisfaction of steering—a thing that always gives me supreme delight. I have a passion for steering boats in a good breeze, and we had one. The only alloy to Talbot and myself was ——'s stupidity with the sailors; he gave them so much good advice!' On the 18th, having reached St. David's, he announced to his wife that 'no guests being expected, the provision was rather scarce, consisting of four very brown and dry mutton-chops. However, with the help of two apple-dumplings, some cheese, bread, and butter, we felt that we had in truth nothing to complain of. It was dark when we got into the town, but I recognised the old houses, especially the first on the road, Captain Propert's, who in old days took the first French prisoner when Pembrokeshire was invaded [in 1797], and tying him to his saddle-bow, rode with him to Haverfordwest. This morning [Sunday] after breakfast I walked down to the Cathedral, and sat in my old stall on the right of the bass vicar-choral, and sang bass just as I used to do fourteen and a half years ago. Mr. Richards preached. He was then a curate, and is now a canon. He was the only clergyman in the church. The congregation consisted of about ten or a dozen women, four or five men, and some boys. The church looked so old, older than any church you ever saw, and though something has been done to it, it still looks sadly neglected. . . . When I came here I had just entered the Survey a month or two before. Sir Henry sent me alone to this extreme corner of South Wales. Except one or two slight acquaintances in the English part of Pembrokeshire, I knew no one in Wales then. Truly, I did not see in a vision that in eleven years I should progress from this igneous and Cambrian county to the extreme northern igneous and Cambrian county of Wales, and there find my mate. I was then twenty-seven, and thought every day a holiday, and nothing about marriage at all, except that I thought if a man were rich enough it would be better to be married than single.

' From St. David's I went to Fishguard, and stayed there all the winter of 1841 and spring of 1842. *Mr.* De la Beche, Kendall and Rose, and Aveline and Rees were all at Fishguard till the autumn of 1841.

In the spring of 1842 I joined *Sir Henry* De la
Beche at Caermarthen after three weeks spent *en
route* at Cardigan and Newcastle Emlyn. From
Caermarthen in the same summer Sir H. went to
Llangadoc, and I to Llandeilo to join little James,
whose bones are now bleaching in the deserts of
Australia. From thence Sir H. and I went together to
Llandovery. In a fortnight I was sent to Pumpsaint,
when Johnes called on me next day to invite me to
a picnic given in one of the caves of the Gogofau.
There was a ball in the evening. That autumn I
spent at Ross and Mitcheldean, and returned to
Dolaucothi at Christmas. I think, were I to go on,
the association of ideas would carry me on all through
my life up till the day of our marriage, which (except
some that have succeeded) was, I think, the best
day in it, for (as I know, and by the consent of my
relatives) it procured me the dearest little wife in
Christendom.'

As the Survey was now creeping eastward across
the southern English counties, the Local Director
could compare the scenery and associations of that
district with those of the more ancient rocks of other
parts of the country, and send home descriptions to
his wife. Thus, while travelling through southern
Hampshire with Bristow, he writes : ' What a striking country we came through to-day to an eye like
mine, which delights in raking up images of the past !
Far-spreading brown heather and moors, with little
mosses and marshes and marshy-banked streams,
broken up with grassy swells covered with native
oaks and other trees, make a true piece of old
England. Wasn't it William Rufus who is said to
have laid all the New Forest waste and depopulated

it to turn it into a hunting ground? I scarcely believe that, in its full extent, for the district forms as a whole one of the worst soils in England, and it is not likely ever to have been thickly populated in these early times. I think its villainous flint-gravelly soil is the reason why it has remained forest-land so long.'

From Lewes, in Sussex, he wrote: 'It is here, you know, that the Russian prisoners are. I have not seen any of them. The soldiers are kept in prison, but allowed to walk out under guard. The officers live in lodgings about the town. I have not told you what a beautiful town it is, clean and airy, such as you see nowhere out of England. The houses are built of brick or of chalk-flints. The streets are hilly, and gardens and trees are scattered about the town. A grey old castle, built by the son of William the Conqueror, stands in the middle of it. The River Ouse runs through the town. The surrounding valleys are full of pretty hamlets, snug farms, and well-grown trees, and the sweeping, green, bare chalk-downs swell all around, from the tops of which (800 feet) your eye ranges far across the lower undulations of the Weald of Sussex to the northern escarpment of the Chalk hills twenty miles off.'

The preparation of the great Welsh Memoir was still Ramsay's chief indoor employment. During wet weather in summer-time, when field-work was impossible, he sat down resolutely to his note-books, maps, and manuscript. In winter he was able to work more continuously on the subject. With the view of securing undisturbed quiet, so unattainable in London, he used to take quarters with his wife and children in some place where one of the surveyors was at work, so that he might have the relaxation of an occasional day in the field. Thus part of the winter of 1855-56 was spent at Cheltenham, where Mr. E. Hull was at that time stationed, and where Ramsay saw much of his valued friend, Thomas Wright, so widely known for his admirable labours among the Jurassic echinoderms and ammonites. In later years he pitched his autumn camp in Scotland.

Among the colleagues with whom he had to consult continually and in great detail was J. W. Salter, whose remarkable knowledge of Palæozoic fossils was of essential service in working out the stratigraphy. But with all his knowledge, it was not always easy to obtain from this palæontologist the definite information which the field-men required. In particular, there was at this time a struggle to get him to draw up tables or lists of the fossils actually named from each locality and horizon. Without these it was clearly impossible to make progress either with the revision of the field-work, or with the preparation of the Memoir. Many were the remonstrances and entreaties addressed to him by Ramsay on the subject. The following letter may serve as a sample :—

CHELTENHAM, 30*th January* 1856.

My DEAR SALTER—I shall be up to the Anniversary, and shall hear your paper.

I have not lost my interest in the South Wales question. Quite the reverse ; for it occupies most of my Silurian thoughts. I have thought much over it, and latterly talked over it with Talbot, and formed my conclusions, which, in many respects, are not dissimilar from yours. But I do not consider my conclusions yet conclusive, nor do I yours, nor Talbot's, and he knows more about it than any of us. However, we are in a fair way. Lists, Lists, Lists, are what we want, and what you want, and without lists the fight is not two-thirds done. You nearly shook us about these Bwlch Trebannon beds being Upper; but not quite, for we could not reconcile it to our consciences that they could be anything but Lower. However, by help of lists and physics, we'll purge the whole question, and have it all straight next spring. As to unconformities I say nothing, and wink my mental eye. Therefore let us have lists. I shall be up in a week or eight days.—Ever sincerely,   A. C. RAMSAY.

In London it was difficult to carry on continuous literary labour, so many colleagues had questions to ask, and so many callers were desirous of a chat or of information. Nevertheless, he contrived to send off to the editor of the *Edinburgh New Philosophical Journal* a review of the fifth edition of Lyell's *Manual of Elementary Geology*. This article appeared anonymously, but its authorship was manifest. It was obviously written by a man of wide practical acquaintance with geology, who could speak familiarly of the geological features of many parts of the British Isles of which no account had yet been published; who could appeal forcibly to evidence of glaciation in central Scotland, mentioning localities that had never been cited before; who could refer to places all over Wales from Anglesey to Pembrokeshire, some of which no geologist out of the Survey had ever visited; who knew Charnwood Forest, had rambled across Shropshire, Cheshire, Staffordshire, and Derbyshire, had extended his observations into Switzerland, and had studied the drifts of the Oberland. There was only one man in Britain who had this range of personal experience, and that was A. C. Ramsay. Lyell had no difficulty in at once recognising the writer. At a party at Dr. Fitton's on the 2nd April Ramsay met Lady Lyell. 'She told me,' he says, 'that to-day Sir Charles had received a review of his *Elements* which bore internal evidence of its authorship, and which, he said, was the best thing that had been done, being the only good review that had ever appeared of any of his works since Poulett Scrope wrote one.'[1] So pleased was Lyell that a few days afterwards he sent the subjoined letter :—

53 HARLEY STREET, LONDON,
6th April 1856.

MY DEAR RAMSAY—I have had time since I saw you to read over more carefully your article in the *New Phil.*, which gives me much pleasure, independently of what is said of my book, for it is no small satisfaction to find a younger man of wide experience, and one who has explored different regions, arriving at similar conclusions on many theoretical points still controverted here, and more so on the continent.

The opinion at p. 313, that neither the Silurian nor Cambrian rocks show traces of a beginning, is one of those useful confessions of faith; also, that the greater the age of a formation, the less chance is there of its deltas being preserved, p. 314; also that the present margins of old formations are the result of denudation, *ibid.*, etc.; but, above all, p. 306, Titanic agency, etc.

I agree with Dr. Hooker that this article in its style is extremely good, apart from scientific depth. He says it is so thoroughly *English*. But for that matter I always maintain that your first paper on Arran left nothing to be desired.

The small number of the fresh-water formations prior to the Tertiary cannot be too much insisted upon, and you have brought it out well. The 'Letten-kohle' of the Trias near Stuttgart is a near approach to an estuarine deposit, to say no more; and it is

---

[1] Alluding to the article written by G. P. Scrope which appeared in the *Quarterly Review* for April 1835.

near that coaly deposit that the *Microlestes* or Triassic mammifer was found in a bone-bed, the exact circumstances of which I examined last year, and will tell you about some day.

Your O opposite the 'Upper Cretaceous' is too often forgotten when the progressive development advocates reason on the absence of Upper Cretaceous mammalia.

In regard to the Carboniferous delta, I would rather accept the idea of many contemporaneous rivers debouching in one sea, than suppose the coal-fields of the United States to have been all once united, as H. D. Rogers supposes.   Assuming these coal-fields to be deltas, it is no doubt strange that we have so few terrestrial fossils; that it should have been left for me and Dawson to discover the first land shell, and in America the first reptilian bones.   Land snakes may have existed on the then continents, even without offending against the laws of progressive development, but when we find them, and helices, and other signs of land creatures, the time will come for speculating on the absence of higher vertebrata.

The four O, O, O, O, or cyphers, opposite the four lowest Palæozoic groups, are significant.   It was also well to insist on the numerous subdivisions of the Oolitic period, and of others, each separately equal to the 'Glacial and recent epochs.'

If I did not take for granted that the condensed essay on the glacial phenomena of Wales and other parts of the world was to appear even more full and expanded in your 'Survey Memoirs,' I should grudge its being given in an anonymous shape to any scientific journal.   The distinctness of the molluscan fauna on the opposite shores of the Isthmus of Panama is well adverted to, and I suspect unanswerable.   Notwithstanding the upraised marine deposits in the N. Polar regions, there is much low land, and so much sea, that we have only to suppose a few such peaks as now lift themselves up in the Antarctic regions (Mount Erebus, etc.), and I believe we have the required cold.

The quantity of reading and original observation adduced, quite naturally and without parade, in the last eight or nine pages, is prodigious, and not more than one in a hundred of the readers of the *Mag.* will know how to appreciate.   The few will do justice to it. I am glad you paid a passing tribute to the 'illustrious' one who boldly led the way against the ridicule and scepticism of the ordinary crowd in regard to Welsh glaciers.

I am rather afraid, I confess, of D. Sharpe's paper, although in the *Elements* I led the way before Murchison in transporting Alpine blocks to the Jura by floating ice.   But D. S. requires twice the upheaval that I asked for as having occurred since the Glacial Period.   To raise mountains 9000 feet has probably required more than a part of one geological period, and Studer and others have in vain looked for marine shells in the Swiss drift.   Fresh-water remains and extinct mammalia I believe they have discovered near Berne, but never at very high levels.

But I must conclude with thanking you for what would have been a treat had the *Manual* of Phillips instead of my own been the subject of your Essay. It has scarcely ever, in the course of twenty years, been my lot to be reviewed by writers who had any practical experience as original observers in the field, and I therefore value your criticism the more.—Ever truly yours,   CHA. LYELL.

At the Anniversary Survey dinner in the spring of 1856, the first presided over by Murchison, who entered heartily into the merriment of the evening, Ramsay produced three new songs and a glee on geological topics. One of the subjects selected by him was his Permian boulder-clay, to which reference has already been made. The style of the composition may be inferred from one or two verses.

> Few, few believe what I have told,
> Men say that I am overbold.
> What then ? they sneered that Welshmen's tails
> Had polished Buckland's rocks in Wales.
>
> And when I'm dead, and these poor bones
> Lie underneath the turf and stones,
> The home of worms and churchyard mice,
> Men then will swallow Permian ice.
>
> Then, then I trust the old Survey,
> Young hands and these, my friends, grown grey,
> Will rear above my mouldering bones
> Four monstrous Permian boulder-stones.
>
> And, on a slab by ice worn smooth,
> Record that in their early youth
> The poor old boy beneath that lies
> Loved well to walk and talk on ice.

The best of the Survey songs ever written by Ramsay was one which he produced next year (1857), and which may be conveniently inserted here. It refers to the geological expedition made some years

R

previously by Murchison, Keyserling, and De Verneuil
across Russia to the Ural Mountains, and was such a
favourite ditty at the Survey anniversaries that its
author was often asked to sing it.

THE LAY OF SIR RODERICK THE BOLD AND THE EMPEROR OF
ALL THE RUSSIAS

AIR—' *The Auld Wife ayont the Fire* '

The auld rocks ayont the sea,
The auld rocks ayont the sea,
The auld rocks ayont the sea,
    That rise upon the Ural.
There was a doughty Scottish knight,
A hammerman o' mickle might,
The Laird o' Taradale he hight,
    Gaed singing ' Tooralooral,
The auld rocks ayont the sea,
The Russian rocks ayont the sea,
I'll map the rocks ayont the sea,
    That rise upon the Ural.'

To Petersburg the knight he gaed ;
The Czar cam down and to him said,
' Ye're welcome here to mak a raid,
    Out ower as far's the Ural.
Frae west to east in ilka hole
Ye'll cast an ee, and 'twill be droll
But you will find a bed o' coal,
    And I'll sing Tooralooral.
The knight he cam across the sea,
The Scottish knight cam ower the sea,
To whack the Russian rocks for me,
    Right out across the Ural.'

The knight he took a working squad—
De Verneuil and another lad,
Count Keyserling, and scoured like mad
    (All singing Tooralooral)—
Silurian rocks and guid Auld Red,
Wi' fish and shells baith in ae bed,
And Permian strata overhead,
    Right up against the Ural—
These auld rocks ayont the sea,
Wi' Oxford Clay ayont the sea,
Erratics o' the Glacial Sea,
    Choke up against the Ural.

Then hame he cam, and left his mates,
And wrote a book wi' maps and plates,
And sections o' the Russian states
    Frae Baltic Sea to Ural.
The Emperor he scratched his poll,—
''Tis bravely done! but by my soul!!
I wish we had some beds o' coal!!!
    Oh! Tooralooralooral!!!!
There's auld rocks ayont the sea,
There's British rocks ayont the sea
Hae lots o' coal, the worse for me,
    There's nane beside the Ural!'

                                            (*Weeps*).

In the summer of 1856 Ramsay had an opportunity of seeing his chief in the field, for they spent
some weeks together in a series of geological peregrinations in the west of England and the borders of
Wales. He thus notes his impressions to Mrs.
Ramsay: 'I am very glad to have been with Sir
Roderick, and to have seen him for many days in his
genuine field-phase, of which I had no idea before.
The impression he makes is most favourable. First,
he is a very early bird; and secondly, he is always so
good-humoured, mirthful, and almost boyish, for
though awfully apt to deliver lectures and to talk far
too much geology, yet he will often tell many queer
anecdotes, and sometimes even talk nonsense. Another thing you will highly approve of—he is always
anxious to get letters from his wife, and very frequent
in his letters to her. I am certain they are very fond
of each other.'

After attending the Cheltenham meeting of the
British Association in August of this year, where
he was President of the Geological Section, Ramsay
took his wife and family to Scotland, and settled for
some months in a cottage at Dalkeith, in Midlothian,
where he could work at the Welsh Memoir and have

an occasional day in the field with his colleagues, Mr. H. H. Howell and the writer of this biography,[1] who at that time, and for some years later, constituted the whole staff of the Survey in Scotland. There could not be a more charming companion in the field than he. So long as the surveyor was untrained the Director would spare no pains in going over his mapping, sometimes spending almost as much time in the inspection as had been occupied in the original survey, and never resting satisfied until he saw that the structure of the country was adequately grasped and correctly mapped. In such educational visits the day passed almost wholly in geological conversation and discussion. But when once he recognised that his subordinate was a careful and conscientious worker and could be trusted, his confidence in him showed itself in many agreeable ways. He would then touch lightly on details, contenting himself with a look at some of the more important sections, and getting a clear notion of the general structure. He would launch out into disquisitions on theoretical questions, more especially on those which had recently been engaging his attention, and would astonish his young companion by the mass of information which he displayed, much of it not obtainable from books. He had a singular gift of conversation, which enabled him to draw out of a man who had any special knowledge to impart such information as served to elucidate geological questions. He not only ransacked books of travel, but he questioned the men themselves who had travelled, and stored up in his memory the facts,

---

[1] I had joined the staff in October 1855, and after some months of field-work with Mr. Howell in continuing the Director's mapping in East Lothian, began the survey of Midlothian in 1856.

allusions, or suggestions which he thus obtained. It was this wide range of knowledge and broad view of geological principles which gave so much interest and value to his lectures, and which made the long talks with him in the field so inexpressibly instructive to those who were privileged to take part in them.

But where Ramsay met with a congenial spirit in those country rambles, geology formed only a part, and sometimes, if the rocks were not particularly difficult or attractive, only a small part of the conversation of the day. English literature was to him a vast and exhaustless garden, full of alleys green and sunny arbours, where from boyhood he had been wont to spend many a delightful hour. When he found among his colleagues one whose talk was not always of stones, but who had ranged like himself far and wide in literary fields, he opened out his inner soul, and his conversation glowed with an animation and power as well as a gleeful exuberance which astonished and charmed his companion. As the writer pens these lines, he recalls many a happy day spent with his Director among the hills and valleys of southern Scotland, when the discussion of the geology and the mapping were interspersed with endless comments on favourite authors, disquisitions on style, analyses of characters in fiction, and quotations of parallel passages in illustration of some thought that had arisen in the course of the talk. Ramsay had his favourite authors, for whom his affection increased every year. He knew Shakespeare so well that he would every now and then flash some apposite phrase or line from him to lighten up the sentence he was uttering. Among novelists his acquaintance was wide and varied, but he always put Scott far away at the head of them all. He had read the *Waverley Novels* so often, and remembered them so vividly, that their characters served in his memory as personages whom he could almost believe that he had actually known.   How earnest he would grow as he discoursed on the plot of *Ivanhoe* and its relation to the known historical events of the time to which it was referred!   His extensive knowledge of English scenery enabled him to picture vividly the surroundings of Gurth and Wamba, and to enlarge on Scott's marvellous power of seizing the dominant features and local characteristics of a landscape, which perhaps he had only seen casually, and straightway colouring them with a vivid glow of human interest. Among English poets one of his greatest favourites was Keats.   He would sometimes speak as if he would rather have been the author of *Hyperion* than of any other poem in the language.   The quaint conceits in the *Ode to a Grecian Urn* delighted him, and various lines in it were often on his lips—'for ever piping songs for ever new,' 'heard melodies are sweet, but those unheard are sweeter,' 'beauty is truth, truth beauty.'   How often have we been requested to proceed with a statement by the quotation, ' Therefore, ye soft pipes, play on !'

There was yet another feature of Ramsay's mind which made these excursions singularly pleasant. Though not in any sense an antiquary, he knew a good deal about the history of architecture, and as has already been remarked, took a keen delight in visiting ruins and trying to form a mental picture of what they must have been before the gnawing tooth of time had dismantled them.   Whatever, indeed, linked him with the past had a charm for him.   He never willingly missed an opportunity of seeing a ruined castle or keep, a mouldering abbey, a grass-grown encampment, or a lonely cairn. If tradition or song invested any spot with a living interest, he would not consider his geological inspection complete if it had not included a visit to that site.

In Scotland, where so many ruins are scattered over the southern counties, and where tradition, legend, and ballad have given celebrity to so many localities, there was during the annual visits of inspection abundant opportunity afforded to the Local Director to indulge to the full his love of the memorials of a vanished past. His letters, written usually either when more or less wearied after a long day of walking, or when hurried in the morning by the preparations for the start, give only a faint glimpse of the enthusiasm which these old associations kindled in him on the ground, as, indeed, they convey a most imperfect picture of the bright sparkle and vivacious earnestness of his best conversation. The wild lonely tract of Lammermuir which he had partially explored from Dunbar afforded him exhaustless materials for indulging in his antiquarian tastes. In some parts of the district every prominent eminence has its circular earthen grass-covered ramparts of ancient Celtic forts, like the Maiden Castle, where Marmion had the nocturnal encounter with the Palmer. Legends still tell of foray and feud, and tradition faithfully points out the scenes of the incidents. After a ramble with the writer through this region, Ramsay wrote to his wife (27th September 1859): 'We have had to drive eighteen miles by a hilly road from Lauder to Duns, and to do as much work as we could upon the hills by the way. We came across a fine wild country growing into cultivation between the southern slopes of the Lammer-

muirs and the English border. We passed a hill-top
on the left, crowned with two cairns, where two
brothers met and slew each other, unknowing who they
were. And far away lay the dark field of Flodden,
where Scotland bit the dust.

'Grieved was I to hear of Dr. Nichol's death.
He was my first scientific friend when growing into
man's estate, and but for him I might never have been
such as I am.'

On the west side of Scotland historical and traditional associations enlivened not less the inspecting
tours of the Local Director. In Ayrshire he could
enjoy himself to the full. The geology had so much
variety and interest that it furnished ample material
for the most solid talk. The scenery embraced the
hills of Kyle and Carrick, with the deep ravines of the
Ayr and the Doon, while westward the whole
panorama of the Arran mountains rose out of the
blue firth. The light of song glowed all over the
region. Every parish had its old castles, its legends,
and traditions. I have never seen my friend more
thoroughly happy than when he was rambling over
that part of Scotland. Each day was full of new
surprises and delights for him. In the morning we
might be tracing out the sites of old Permian volcanoes,
or following the succession of lava-sheets in the Old
Red Sandstone, but before evening we were pretty
sure to get into old ballads and traditions, suggested
by the associations of the localities through which our
work led us. The old castle of Auchendrane, perched
so picturesquely in the ravine of the Doon, with its
charming family circle and its hospitable host, so fine
a sample of the antique world, filled him with raptur-
ous delight, and formed many a time in later years the

subject of his thoughts and his conversation.[1] He was a welcome guest too at Dalquharran Castle, in the Girvan valley, where his love of the past was amply gratified, where his host could retail many reminiscences of men, manners, and customs that had long passed away, and where his hostess threw over the household the inexpressible charm of her own gentle and gracious presence.

In the eastern part of the district where the infant Nith turns southward into the dale which bears its name, he was much struck with these local associations, as may be seen in the following passages from a letter to Miss Johnes : ' The low bit of country in which I now write [New Cumnock] is 700 feet above the sea. "The wind is howling in turret and tree," or would do so if there were either turrets or trees here for it to howl in. We are in a great plain, once a lake, now filled with peat-moss, through which slowly flows the winding Nith near its sources. Great rounded green hills rise all around, some of them more than 2000 feet in height, and beyond and among them are vast moors and mosses, swelling and undulating for miles and miles, amid which " the hill-folk " took refuge in the days of Claverhouse and the " bluidy Dalzell." Two days ago I was at Kirkconnel, on the Nith, some six miles from here. Old Kirkconnel, now desolate, is but a ruined church, mere foundations, with a churchyard full of mouldering tombstones, at the foot of the desolate hills. I searched them all, and removed the grass and moss from some, in hopes of finding thereon "Hic jacet Adamus Fleming," but in vain.[2]

[1] Auchendrane was the prototype of a sketch of ' Balbraith ' which he afterwards wrote for *The Saturday Review.*
[2] There is another Kirkconnel farther south, on the Kirtle Water, which claims to be the true scene of the ballad, and where Helen's grave is pointed out

'I wish I were where Helen lies,
    Night and day on me she cries;
  O that I were where Helen lies,
    On fair Kirkconnel Lea!

'Another theme of interest in these parts to human mortals is the aforesaid wide dreary mosses, where in old times the steadfast covenanting hill-folk used to bide the weather with Bibles in their bosoms when fleeing from Claverhouse, Dalzell, and the rampaging dragoons. Many a spot where after "testifying" they were shot and buried in the moss has been pointed out to me in my geological rambles, for even geologists have human interests.'

Until the year 1867, when the Geological Survey of Great Britain was divided into two, and a separate establishment was given to Scotland, Ramsay used to visit his staff in the north each year, inspecting the field-work, and enjoying a renewal of his acquaintance with his native country. After that year his visits were few, and ceased to have any inspecting duty connected with them.

with a decayed trunk of an ancient thorn rising above it, from the roots of which a young sapling has sprung up.

CHAPTER VIII

DURING the six years from 1857 to 1862 Sir Andrew Ramsay spent a part of each summer but one abroad. It was in this part of his life that he accomplished almost all the work in foreign geology which he ever did. There will therefore be some convenience in the treatment of the subject if we group the tours together in one chapter.

From what has been stated in the foregoing pages it will be clear that the ice-fever in geology had now got full hold of him. He had seen a little of Swiss glaciers, but not nearly enough to enable him to answer all the questions which the glacial phenomena in Britain were continually putting to him. He therefore determined to devote as much time as he could spare to the study of ice and its work outside the narrow limits of the British Isles, while neglecting no opportunity of investigating the subject within these limits. By a happy accident he was soon able to carry out this determination in a far fuller manner than he could have dreamed to be possible. How this came about is told in a letter to his brother William. 'Certain of the great steam-boat companies, at the solicitation of the Canadians, have put a few free passages to America and back at the disposal of the leading scientific societies to [enable delegates] to attend a meeting of the "American Association for the Advancement of Science," which takes place at Montreal on the 12th August [1857]. Failing Sir Roderick, the Geological Society have deputed me to represent them, so I go in an honourable position.' Taking Mrs. Ramsay with him, he sailed on the 29th July.

This was by far the most enjoyable and instructive of all his foreign expeditions. His friends, Sir William Logan and Professor James Hall of Albany, spared no pains to ensure his seeing everything that he wished to see, or which they thought it important from a geological point of view that he should visit. His time was thus economised to the utmost. He was taken from point to point, so that in the course of exactly two months he had travelled over a large extent of country, and had been conveyed over those tracts which were specially of service to him in reference to the problems in which he took interest.

From Montreal Logan carried the two travellers to Ottawa and up the St. Lawrence by the Thousand Isles and Lake Ontario to Niagara, thence to Lake Huron. At Sarnia, Hall met them, and brought them into New York State by Genisee to his hospitable home at Albany, from which centre they made excursions to Schoharie, the Helderberg, and Catskill Mountains. Descending the Hudson to New York, they found there that almost all the persons to whom they had introductions were absent on holiday. They therefore passed on to Newhaven, and paid a short visit to Professors Dana, Silliman, and Brush, thence to Boston, where they were delighted to meet Agassiz, and so back to Montreal and Quebec for the voyage home.

The chief geological fruits of this expedition were given partly in a discourse to the Royal Institution, but more fully in a paper read before the Geological Society. Ramsay had not yet realised the massiveness of the land-ice of the Glacial Period. Like most of the geologists of the day, he still regarded the 'drift' as the result of transport by icebergs, and to the same agency he attributed the striæ on the sides and summits of the hills. He recognised the remarkably ice-worn character of Canadian topography, but he did not yet associate that character with a former extensive glaciation by land-ice. Nevertheless he now beheld the effects of this glaciation on a far grander scale than he had ever before seen them, and unconsciously he was accumulating material that would enable him to get rid of the paralysing idea that the land must have been submerged beneath the ocean as far as the highest striations or drift deposits could be traced. He was not, however, able entirely to divest himself of the old error until the summer of 1861.

In the summer of 1858 Ramsay and Tyndall made an expedition together into Switzerland for the purpose of studying the phenomena of glaciers and ice-action. The results of their conjoint observations on this occasion are to be found in the writings of each explorer. Tyndall had not specially examined the proofs of the former greater extension of the glaciers of the Alps, and Ramsay, to whom this was a matter of supreme interest in connection with his investigations in Britain, took pains to direct his companion's attention to the subject during the course of the excursion. Arriving at Grindelwald, they undertook some preliminary climbing among the ice-filled valleys of that district, and Ramsay proved

himself so expert a pedestrian that Christian Lauener
deemed it quite practicable to proceed on the proposed
series of ascents with only himself as guide. They
crossed by the Strahleck Pass over to the Finsteraar
and Unteraar glaciers, spent some time at the Grimsel
studying the marvellous evidence of the vast dimen-
sions of the ancient Alpine ice, went to the Rhône
glacier, and then to Viesch, the Æggischorn and the
Märjelen See, where they remained some days taking
measurements of the thickness of the ice and the depth
of the glacier-lake, and making observations of the
temperature of the air. Ramsay had here the great
satisfaction of watching the origin and movements of
icebergs. Descending the Rhône valley to Visp, they
walked up to Zermatt with the intention of ascending
Monte Rosa. The ample details of geological observations in Ramsay's note-book of these rambles were
afterwards condensed by him in his essay on the Old
Glaciers of Switzerland and Wales. They show how
continually his experience in Britain enabled him to
interpret the phenomena in the Alps, and, on the other
hand, how the existing snow-fields and glaciers of the
Alps gave new clearness to his conceptions of the
vanished ice-sheets of his native country. One or two
citations from the note-book may fittingly find a place
here.

'*Viesch, 30th July.*—Started at nine for the Æggischorn. On the partial clearing of the mist, ascended
the mountain. Tyndall and Lauener pushed on before
me, and were at the top twenty minutes or so earlier
than I was. The day is not far past when I was at
least a match for either of them. Tyndall cannot
believe that at forty-four and a half years my best
days, as regards strength and agility, should be gone,

and he makes no allowance for my having reached the top of the curve, and begun to descend on the other side.

'The summit of the peak consists of piled blocks of gneissic rocks, rent by frost and weather, and heaped on each other in wild confusion, like the summits of the Glyders, or Y Tryfan, above the passes of Nant Francon and Llanberis. The view from the summit was, indeed, grand. Below on the north and west lay the Great Aletsch glacier, seemingly as much larger than all the other glaciers I have yet seen, as the St. Lawrence is larger than the Severn, Thames, or Seine. There it lay below us, broad, smooth, and sweeping, although crevassed and somewhat crumpled. The moraines looked small upon it. On the left it descended into the valley, and on the north it was lost in the far recesses of those Alpine giants, the Jungfrau, Mönch, and Finsteraarhorn. On the north-west the glacier is joined by two great tributaries, one the Ober Aletsch glacier, the other the Middle Aletsch glacier, stretching up among the snows and awful cliffs of the Aletschhorn, the peak of which rises more than 13,000 feet above the sea. This summit is higher than the Jungfrau. White sunny mists were seething round it, half veiling and adding to its majesty.

'Seemingly close below lay the Märjelen See, with the glacier branching into it, and breaking off in large masses, which floated away eastward as tabular bergs before the wind, and grounded on the desolate shores. Clearly the glacier once sent off a branch down this valley, for besides that it partially does so still, the rocks on the hills by the lake are *moutonnées* high up on either side. The glacier must then have sent off a branch that united with the Viescher glacier, and at a

certain period of its history it sought the valley of the
Rhône by two channels, that of the Aletsch and that of
the Viescher glacier.'

On reaching Zermatt he found letters telling him
that his mother had had a serious attack of bronchitis,
but had somewhat rallied. Next day he made the
following entry in his note-book :—

'*Zermatt, 9th August.*—Found at the post office a
black-edged envelope, which at once told me that my
mother was dead. I merely read the first few lines,
and then ran up the mountain after Tyndall, towards
the Riffel Hotel, but he had gone to the end of the
Gorner glacier, and I outstripped him. When half-
way up, exhausted with my speed, I turned and saw
two figures far below by the glacier, whom I guessed
to be Tyndall and Lauener. During the half-hour
they took to come up to me I had leisure to read my
wife's letter, and my grief found a little vent. Tyndall
came up, and I marched down to him with my hat
drawn over my eyes. We arranged that Lauener
should go down and countermand the guide, who
next day was to accompany me up Monte Rosa, and
Tyndall persuaded me that instead of starting so late,
it would be better to remain with him and go next
day. So we ascended to the Riffel.'

He started homeward early next day, and walked
the rough thirty miles of valley down to Visp to
regain his portmanteau, and catch the diligence for
Bex. Finding when there that he could reach
London almost as soon by sleeping at Bex as by
going on to Geneva, he remained to have a look
at the famous blocks of Monthey. He 'wandered
among them half a summer's day, pleased and amazed
by their beauty and great size, and the evidence of

power conveyed to the mind while reflecting on the agency that bore these ponderous masses and left them perched on this hill, from 500 to 600 feet above the Rhône. The largest, twenty-two paces in length, and nearly equally broad and high, has on its flat summit a good-sized summer-house with a small garden containing cherry-trees.' [1]

On reaching England, and realising there amid all the old familiar surroundings the blank that had now fallen upon his life, with the rupture of his oldest and tenderest associations, he made the following entry in his diary :—

'On the 29th July 1858 my dearest mother died at the Bridge of Allan. She had been a few days ailing, a little breathless, and in bed. William had gone to Glasgow, and was telegraphed for ; when he got back at six o'clock all was over. There may have been many as good, but none better than our mother. She died in her eighty-fifth year, surrounded by love. She truly lived all her days, in health and cheerfulness, in peace, love, and honour, with her faculties and cheerfulness clear to the last, loving books and mirth, and writing a good letter in a clear hand to the very end. When my father died she must have been fifty-three years old. I was then thirteen. She had but £1000 and a house. Then came a time that would have crushed a weaker spirit. But she battled for us, and keeping college and other boarders, brought us all up respectably. William was apprenticed to Napier, the engineer, and I went at that early age into ——'s counting-house, and passed through many battles ere I emerged from mercantile life and got launched in the world of science. These times, which I look on

[1] *Old Glaciers of Switzerland and Wales*, p. 30.

as hard, though I was when young merry enough, my mother never grumbled at, but doing a duty was happy in it, and when we began to do well and made her give it up, she almost missed for a time the employment to which she had been used for eighteen years.

'Ere her death she had thirteen years of peace and quietness, and every year endeared her more to those who knew her best. My wife loved her like a veritable daughter, and all the children that approached her loved her also. Her memory is so pleasant to me, and all her deeds, her courage, kindness, charity, and goodness; she lived her time in the world so well, and so completely fulfilled a good woman's mission, that though I miss her, and every now and then think "I must write to my mother," yet my sorrow is tempered by a thousand pleasant reflections. She lived to the last happy and contented, beloved by all, happy in all her children, and she scarcely seemed to die, so easy was it to pass from one world to another.'

On returning to England from this Alpine excursion Ramsay had to address himself to a long course of arduous labour. Partly from the necessities of his official position, and partly from his own voluntary act in undertaking various pieces of work outside the claims of the Survey, he was now involved in a greater pressure of mental toil and accompanying worry than had ever befallen him before. The inspecting duty in the field was every year becoming more exacting, as the staff of officers increased and the area of survey augmented. But had that been his chief or only occupation, he would have made it in some measure a kind of holiday employment. But there was now a large and growing amount of literary work thrown upon him which was unknown in the older days of the Survey. Sir Roderick Murchison had arranged that each of the one-inch maps, as it was published, should be accompanied with an explanatory memoir, so that the public might be put in possession of the chief data used in the construction of the map, and of the information needful for its proper interpretation. These memoirs were to be edited by the Local Director from the manuscript notes supplied to him by the officers who had surveyed the ground. He sometimes had to furnish additional material from observations of his own, and the amount of editorial supervision was thus often exceedingly heavy. Then the great Memoir on North Wales still dragged its slow length along. From various causes, but chiefly from the want of sufficiently full notes by one or two of his colleagues, Ramsay had been unable to make rapid progress with this large and detailed volume; though it had been for so long his chief indoor employment, and though he again in the autumn of 1858 took a house in Scotland for three months, this time at St. Andrews, in order to push it forward.

Another task occupied some part of his thought and time. He had planned a descriptive catalogue of the rock-specimens in the Survey collection in the Jermyn Street Museum, and while assigning certain portions of it to three of his colleagues, had kept the main share of the work in his own hands. As ultimately published, this volume formed an excellent compendium of British geology. In particular, the account of the successive volcanic episodes in the Palæozoic period in Britain was by far the best which up to that time had appeared, and it was mainly the work of Ramsay himself.

But over and above these official labours his hands were full of work. He at this time condensed the information on the published Survey maps, and produced a geological map of England and Wales on the scale of twelve miles to an inch, which is still the most serviceable general map of the kingdom. He prepared a Friday evening discourse for the Royal Institution on the geological results of his Canadian excursion, and wrote out a fuller statement of the subject for the Geological Society. He drew up for the well-known volume, *Peaks, Passes, and Glaciers*, a chapter on the old glaciers of Switzerland and Wales. This essay, full of original observation, and suffused with the charm of freshness and enthusiasm, is one of the most important and delightful which he ever wrote. It was reprinted as a separate little volume, and has long taken its place among the choice classics of glacial geology. He now began to write for the *Saturday Review*, and for a number of years continued to furnish occasional articles to that journal, chiefly on geological topics, but without the technicalities of the more formal communications to learned bodies.

His habit at this time, when in country quarters in the autumn, was to write during every available hour of daylight, and only to go outside for exercise when it was too dark any longer to see his manuscript. In the end the strain proved too great both on his brain and on his eyes. In the summer of 1859 he accompanied Murchison into the North-West Highlands of Scotland, and assisted him in the preparation of his discourse for the British Association at Aberdeen. He seemed tolerably well and merry at that meeting, but afterwards, when out among the hills in the south of

Scotland, he complained of weariness.[1] The symptoms of mental exhaustion increased during the autumn. He had for the first time taken a permanent house in London, having hitherto only occupied furnished rooms. But hardly had he settled in the new home when it became evident that he was in no fit condition for London life, and more particularly to undertake his usual course of lectures at the School of Mines. Towards the end of December it was arranged that the lectures should be given by his colleague Jukes, while Ramsay himself went to the house of his helpful and sympathetic friend, Dr. Wright of Cheltenham, under whose care, with entire cessation of work and worry, it was anticipated that speedy convalescence would be secured. But the recovery was not to be so easily effected. He was ordered to abstain from all work for a time, and in order to obtain complete rest and change, he and Mrs. Ramsay with the children went abroad. He wrote to me just before leaving (31st March 1860): 'I sail on Tuesday with bag, baggage, fishing - basket, rod, flies, sketch-books, Shakespeare, and the musical glasses in the shape of sundry minor authors. I am to be away, if nothing specially intervene, for six months, but I think the less we say about *that* the better; and if any one by any chance asks you anything about it, *several* months is a convenient word, with the addition that I join the Scotch geologists as soon as I come back, and also that Sir Roderick says he will pay

---

[1] During one of these rambles with me in Fife our conversation turned on the Boulder clay and the mysteries of its origin. We both felt how unsatisfactory was the received explanation of iceberg action and submergence. I was thus led to study this deposit, and to reach thereby the conclusion, at which Ramsay also simultaneously and independently arrived from a consideration of other evidence, that the great glaciation was the work of land-ice. This change of view was completed before the summer of 1861.

you a visit during my absence. Gossips have been
exaggerating my illness, and I know that, both on my
own authority and that of the doctor, you will do me
the friendly turn to give a flat, blunt, sharp, plain,
broad, profound, high, and indignant denial to any
statement that I am seriously ill. I am even now so
wonderfully better that I can do a good hard day's
work at the office, albeit I am tired at night, and
therefore to set me up alike by day and by night, an
entire cessation for a while is needed.'

Nearly two months were spent in Bonn, and of
this sojourn Ramsay used always to talk with much
enthusiasm. He loved the great river, and delighted
to sit quietly smoking and watching the 'breast of
waters' as it swelled beneath him. He made some
pleasant friends, among whom he specially counted
Von Dechen, the venerable Nöggerath, and young
Ferdinand Zirkel. Professor Zirkel has sent me a
letter with his reminiscences of Ramsay, which is here
inserted :—

My first meeting with the never-to-be-forgotten Ramsay was in
the spring of 1860, in Bonn, so far as I remember, towards the end
of April or beginning of May. One day my fatherly patron and
official chief, Von Dechen, sent for me and told me that an English
geologist, a man of great importance, had come to Bonn with his
wife, to spend some time there for the sake of his health, and in order to
make some geological excursions in the neighbourhood. As Dechen
himself had not time, I was asked to accompany and guide the
stranger in these rambles. No request could have been more agreeable to me. Although I was then only twenty-two years of age, yet
I knew the nearer and farther environs of my native town as well
as anybody. I was at that time a pupil of the Prussian State-
Mining Institute.

So I waited on Ramsay, who was staying at Ermekeil's Grand
Hotel Royal on the Rhine, and then began a series of blissful days.
Sometimes for the whole day, sometimes in the afternoon only, we
rambled in the Siebengebirge, to the Roderberg, to the Laacher See,
to the Devonian Eifel-limestone at Bensberg, and many other places.

When we got back Ramsay would usually have me to sup with him and his wife in the hotel. I conceived at that time an enthusiastic admiration for Ramsay, both for his amiable, simple, and straight-forward nature, and for his acuteness and his range of acquirements in geological matters. I remember in a gully in the trachyte-tuff he suddenly made a couple of steps forward, exclaiming, 'There is a dyke!' and there, sure enough, was a dyke of solid trachyte, which nobody had ever noticed before in this well-frequented path.

I think Ramsay spent a happy time that season in Bonn. Dechen once gave an evening party in his honour. His intercourse with old Nöggerath would have been greater, had the latter not been utterly ignorant of English. I got on extremely well with Mrs. Ramsay.

In the summer of 1860 I made a journey to Iceland, and as, on my return, I spent a short time in England, I then saw Ramsay again. During the first days of my stay in London he was in the country, but I met him on the last day in the Museum, Jermyn Street, in company with Lyell. I told him a good deal about my tour in Iceland, and he presented me with several books.

In 1868 I once more saw Ramsay, both before and after my visit to Scotland. He lived at that time in Upper Phillimore Place, in Kensington, where I spent an evening, and met Howell and Hull. As I left London he gave me a letter to you, and this letter I presented to you in Largs on Monday, 8th June 1868 (according to my diary). That was the day when I had the good fortune to make the personal acquaintance of my friend Geikie.

The last time I saw Ramsay was at the meeting of the British Association at Sheffield in 1879, when I was staying with Sorby.

From Bonn Ramsay and his family moved up the Rhine, and then ascended the Moselle to Alf and Bertrich. There he established himself for a while, and spent his time fishing in the river, exploring the Eifel volcanoes, and gazing with ever-increasing interest upon the great tableland and the valleys cut out of it by the Moselle and its tributaries. From that quiet life he journeyed to Treves, then back to Heidelberg, and into the Black Forest. He attended the tercentenary celebration of Basle University, and even got as far as Munich. But in October he was once more back in England.

The following portion of a letter to Mrs. Cookman (4th July) from Bertrich gives a picture of how the time passed there. 'Having stayed seven weeks at Bonn, and excursed and fished, we steamed up the Rhine to Coblenz, slept one night there, and next morning steamed up the Moselle to Alf, where we remained a fortnight, walking and idling, and fishing again. I assure you I can throw a fly as prettily as need be. But who shall describe the glories of the Moselle with its unutterably tortuous windings, its vineyards, its quaintly-gabled towns, and all its castles, so stately in decay! I am going to buy one for 50 or 100 thalers (£7 : 10s. or £15), and in memory of my late illness, take my title from it—Baron Beilstein.

'Bertrich is a pretty little village, with two or three hotels, baths, and gardens with music in them twice a day. Gaiety there is none, but peace and quiet and a billiard table. The village is set in a deep valley, and three extinct volcanoes crown the tableland above, for hills proper there are none. I have been on foot with a Dutchman (whom I lamed) all over the Eifel, and have seen lots of extinct volcanoes —most interesting. The structure of the country, its physical geography in fact, is most curious—a great tableland, about 1200 feet high, through which the Moselle and other rivers run in deep valleys. On this tableland are perched old volcanoes of Miocene (that is, of Middle Tertiary) age. The valleys are of older date than the volcanoes, for sometimes you see a lava-stream that has run from the mouth of the craters into the valley below. The Devonian strata of the tableland are awfully disturbed, not by the volcanoes, but by far older forces. It was a great plain, so to speak, with valleys scooped out of it long before the lava began to flow.

'We leave this soon, but, before doing so finally, will pay a visit to Treves, to see the northern capital of the Emperor Constantine.'

Though these months on the Continent were spent as far as possible in idleness, Ramsay could hardly find himself face to face with new scenes without being led to notice and reflect on the features in them which bore on any of the questions in geology and physical geography which had always been with him such favourite subjects. His excursions among the Eifel cones and craters gave him fresh material for his work among the old volcanoes of Wales, and for his lectures at the School of Mines. His rambles over the great tableland of that region, and among the streams which have so deeply trenched it, furnished him with illustrations of river-action of which, though he perhaps hardly realised at that time their significance, he was in a few years to make excellent use.

The sojourn in Germany, and the idleness enjoined upon him, had one effect, which was the first to strike the eyes of his friends when he got back to England. He had buried his razors when he left home, and returned with a bushy beard, which he continued to wear during the rest of his life. But though much better in general health than when he went abroad in spring, he was still far from having regained his old vigour and power of work. Indeed, it is doubtful if he ever again was capable of enduring the same mental and physical strain as he had been before his illness. He had again to be assisted in his lectures during the winter,[1] and was still unable for much literary exertion. The Welsh Memoir had to stand aside. Once or twice in the course of the summer of 1861 he amused himself with writing a paper for *The Saturday Review*, including one of the best of his contributions to that journal, on 'Lyell and Tennyson'—an essay which, with its humour, its poetry, its geological aroma, and its literary deftness, is an excellent sample of his fugitive pieces.[2]

Later in the summer he went once more with Mrs. Ramsay to Switzerland for more mountaineering, and to cross over to the Italian side, in order to see the great glacier moraines of Ivrea. The general outline of this expedition is given in a letter to his sister, written from the Stachelberg on the 4th September: 'Ever since we left home we have had perfect weather. We have only had two half rainy days ·in all, and generally there has not been a cloud in the sky. Louisa and I travelled as far as Cologne together without stopping. I then went direct for another day and night to Teplitz, in Bohemia. Thence, after three days' rest and light work, I descended the Elbe to Dresden, across Saxony and Bavaria to Lindau, on the Lake of Constance, and having travelled two days and nights, reached Berne at ten o'clock at night, not a bit tired. Next morning after breakfast I joined Louisa and Mr. and Miss Johnes and Mrs. Cookman at Thun—the most lovely spot in the universe. I stayed there from Friday till Monday, and then left them by steamer on the lakes for Meiringen. There I shouldered my knap-

---

[1] I took this duty, and thus came to have an intimate knowledge of his lecture materials and his methods of preparation and illustration.
[2] *Saturday Review*, 22nd June 1861.

sack at four in the afternoon, and marched alone up
the long, rough valley of the Hasli Thal to the
Grimsel, which I reached well tired at half-past ten.

' I met Tyndall there and some other friends, spent
a day on the Rhône glacier, and ascended the Seidel-
horn alone. Next day, Tyndall not being very well,
I walked to Obergestalen, and the day following
crossed with a guide over a famous pass called the
Ober Aar Joch to the Æggischorn. It took thirteen
hours, ten of which were spent on the ice. The pass
is about 11,500 feet high.

' In the meanwhile Louisa and the party came round
by the Gemmi Pass, great part of which can be done on
horseback, and the second day after my arrival joined
me at the Æggischorn. We took them up a mountain over 10,000 feet high, and on the Great Aletsch
glacier, which is the longest in Europe. Thence
we went to Visp, in the Rhône Valley, and next morning at six they rode and I walked up the valley to
Zermatt, which we reached about six o'clock at night.

' We stayed at Zermatt six days, on one of which
I, with some others, ascended the Lyskamm, 14,891
feet high. There were eight of us, with five guides
and two porters to carry provisions. Having slept
at the Riffelberg, which saves a climb of some 2000
or 3000 feet, we started at twenty minutes to two in
the morning and crossed the Great Gorner glacier by
the light of a full moon. At dawn we were at the
foot of Monte Rosa on the snow, and by 11.40 we
reached the top of the Lyskamm. We went in two
parties, all roped together. The final ascent was
excessively steep, all on snow and ice. Sometimes
we had one leg in Italy and the other in Switzerland.
That part took nearly three hours. The descent

proved nearly as difficult as the ascent, but we all got
back to the Riffelberg by 7 P.M., and down to Zermatt
by a little after nine, having been nearly twenty hours
on foot.   Several previous attempts had been made
to scale this mountain, but all had failed.

'From Zermatt we all crossed the Theodul Pass
(about 11,000 feet) into Italy.   The ladies rode up to
the ice of the glacier, which they reached at seven in the
morning.   They had then three hours walking on the
ice and snow, and by twelve o'clock we were at Breuil,
where we rested and slept.   Next day they rode on
asses to Chatillon, a beautiful old Roman and Italian
town.   Next day with Dr. Sibson we drove to Ivrea,
where we stayed a day, and then on by Chiavasso,
Milan, and the Lake of Como to Lugano, where we
stayed two days, and left the Johnes.   Sibson, Louisa,
and I came across the St. Bernhardino Pass in a
diligence to Hinter Rhein, near the sources of the
Rhine.   There we halted three days, and Sibson and
I scaled two mountains among the glaciers, one of
which took fourteen hours.   Thence we came by the
Via Mala to Glarus and Elms, "did" another splendid
pass, and came on here.   To-morrow is our last day
on the ice; on Friday we shall be in Zurich, and on
Monday evening at our own house in Kensington.'

So far as his physical powers were concerned,
Ramsay seems to have returned to England invi-
gorated by his Alpine exercise.   But he had not regained his old elasticity of mind, and soon began again
to complain of the weariness of work.   Nevertheless,
he braced himself for the duties of the winter, and
succeeded in getting through his lectures to the
students at the School of Mines without help.   He
likewise found himself able at last to sit down to a

congenial task, and to commit to writing the thoughts
and conclusions which had been shaping themselves
in his mind for several years past regarding the origin
of lake-basins.   This problem in physical geography
had never been seriously attacked, and no tenable
solution of it had yet been proposed.   It was his
experience in Canada, and the sight of the lake-
sprinkled surface of the ancient gneiss of that region
which first definitely called Ramsay's attention to
this subject, though he had returned from America
still in the belief that the older and greater glaciation
was accomplished by floating ice during a time of
submergence.   But the recognition to which he had
now come, that that glaciation was the work of the
grinding action of stupendous sheets of land-ice, gave
an entirely new turn to his thoughts regarding the
terrestrial contours of glaciated regions.   In his journeys in Wales, Scotland, and Switzerland he was now
always on the watch for facts bearing on the connection between the traces of ice-movement and the
contours of the ground over which the ice had moved.
He had at last come to the conclusion that the prodigious abundance of lakes in the glaciated regions
of the northern hemisphere could not be accounted for
unless they were connected in some way with ice-
action, and he inferred that in a vast number of cases,
where the lakes lie in rock-basins, these basins have
actually been scooped out by the grinding power of
land-ice.   These observations and inferences he now
proceeded to elaborate as a paper for the Geological
Society.

Before the paper was ready, however, the presidency of the Society was vacant, and there was a
general feeling that it should be offered to Ramsay,

if the state of his health would allow him to accept it.
The President usually confers with former presidents
of the Society in regard to his successor before actually
proposing a name to the Council ; but on this occasion
the President, Leonard Horner, was in Florence, and
unable to take any personal part in the negotiations.   Lyell strongly favoured Ramsay's nomination.
Murchison was afraid of the strain upon his colleague,
if he accepted the duties of this office in addition to
all that he already had to discharge, and urged him
to consult his medical adviser.   On the 3rd February
1862 Ramsay's diary received the following entry :
'Sir R. in a great fuss because I had not seen Haden
[his doctor].   Drove out to Haden's at one o'clock.
He vowed by Jove that he would not stand between
me and the presidency.   So I drove back and told Sir
R., and he said that that settled the matter.'

So at the Anniversary, on the 21st February, he
was duly elected President—an honour well earned by
twenty-one years of continuous devotion to geology,
and the large part taken by him in the work of the
Geological Survey.   In the evening he began his
duties by presiding at the annual dinner of the Society,
where, with the Duke of Argyll on his right, and Lord
Ducie on his left, and most of the leaders of geological
science around him, he had the satisfaction of seeing
a company of nearly ninety assemble to celebrate the
foundation of the oldest geological society.   Those
of that company who still survive will remember the
admirable way in which the new President spoke.
Never before had he so distinguished himself in the
difficult art of post-prandial oratory.   In returning
thanks for his health he showed a quiet dignity and
simplicity, with touches at once of humour and pathos,

which went straight to the hearts of the listeners, and called forth many rounds of applause.

At the very next evening meeting of the Society the President gave his paper on lake-basins. Its conclusions were so startling a novelty in geological physics, and were based on such a mass of detail, requiring careful study, that they could hardly be adequately discussed by an audience which heard them for the first time. Ramsay did not read, but spoke his paper, and being full of the subject did full justice to it. 'Lyell,' as he said afterwards, 'damned the paper with faint praise, and Falconer vigorously opposed it. It was admirably defended by Huxley. The meeting was so lively as to remind us of the old days of Buckland and Sedgwick. Some account of the theory propounded in this paper will be given in a subsequent chapter of this biography. It was attacked by various writers, notably by Lyell, Murchison, Falconer, and Ball, and to some of the onslaughts made on it its author replied in the pages of the *Philosophical Magazine* and *The Reader*.

The following letter, written towards the end of this year, gives a picture of the reception of the paper, and the ferment that arose from it :—

LONDON, *9th December* 1862.

MY DEAR MRS. COOKMAN—By this post I send you the other pamphlet on the origin of Alpine, Welsh, American, Schwartzwald, and Scandinavian lakes. The smaller one I sent you the other day was a pendant to it, and was written *à propos* of a paper by Tyndall in the *Phil. Mag.*,[1] in which he ran the theory of what ice has done to a wild extreme.

[1] November 1862, p. 377.

It was published in *The Times* also, and I thought it
a pity to let it be supposed that my theory led to such
extravagance.

   I do not suppose you will find fault with the
paper on the ground that it wants boldness.   When
it was read Dr. Falconer of Indian-fossil-elephant
celebrity made an onslaught on it of forty minutes.   I
observe that most of the men older than myself re-
pudiate it, while most of the younger bloods accept
it.   Lyell rejects, but then I have Darwin, Hooker,
Sir William Logan, Jukes, and Geikie.   When I explained the theory to Sir William before it was read,
he said : If you don't publish it for America, I will.'

   So strong was the opposition among the older and
more staid fellows of the Geological Society that
Ramsay used to assert that had he not been the
President, and thus in a manner privileged, the
Council would have voted against the publication of
the paper, except in briefest abstract.

   Before the end of the first week in September
1862 Ramsay was glad to escape once more from
London to Switzerland.   There were various geological matters which he longed to investigate more
fully, and as he went this time with only his friend
Dr. Sibson, an accomplished mountaineer, he was free
to arrange his route as the work to be done might
require   Making straight for Geneva, the travellers
first went to Bex, and rambled once more among the
blocks of Monthey.   The weather proved most un-
favourable for mountain-climbing, and after waiting
some days in rain and mist, they resolved to move
into the sunnier clime of the Italian side.   Crossing
by the Sanetsch Pass from Gsteig to Sion, they were

fortunate to find the clouds clearing away. 'Ere we reached the watershed,' he wrote to Mrs. Ramsay, 'there was no mist, except in some of the great corries and up on the highest peaks. The sun shone brightly. We diverged a little from the road to see the end of the Sanetsch glacier. The pass is 7123 Paris feet high. There is therefore no snow on it. While lunching on the moraine we said: "Let us leave our baggage here, and go up the glacier to the Tour de St. Martin and see the great cliffs that over-look the Valais." So at twelve we started, well roped together; but the glacier proved so easy that there was no real occasion for the rope. In two hours and a half we were across the glacier, and saw those noble cliffs 1000 feet and more plump down. We also saw a flock of more than twenty-five chamois not far off, and all the great range across the Valais, from Mont Blanc to Monte Rosa, clouded in places. In an hour and a half we were back at our baggage, and started for Sion at five o'clock. In two hours it was dark, and the guide being nobody, I went ahead, and on a true Swiss road, by torrent and in forest, piloted all safe to Sion by instinct. We got there at half-past ten, having walked fifteen hours.'

Once more in the valley of the Rhone, they ascended to the Bel Alp and the Æggischorn to make further observations on the great Aletsch glacier and its surroundings. Then retracing their steps, they made their way by Turtmann over to the Italian side, and so down the Val d'Aosta to Ivrea, and thence to Turin. Once in the capital of Pied-mont, Ramsay called on his friend Quintino Sella, known abroad as an able geologist, but to the mass of his own countrymen familiar only as their distinguished

Minister of Finance. Of the short time in Turin Mrs. Ramsay received the following pleasant narrative: ' From the post office I went to the Ministry of Finance. The attendant in the ante-room, doubtful of a stranger in a wideawake, said the Minister was engaged with the Minister of Home Affairs, and would be so until late in the evening. I sent in my card, and he came back with a changed countenance and ushered me in. Sella shook me by both hands, and said he was uncommonly glad to see me, and that if I would wait till he wrote a note, he would himself take me to Gastaldi. . . . Gastaldi received me like an old friend, and he has been almost constantly with me ever since. . . .

' I have just come back from the Ministry a *decorated* man, with white and gold cross and green ribbon ! The royal letter and decree are to follow. . . . I leave to-night, and cross Mont Cenis, arresting myself perhaps at Macon for the second night.'

The knighthood thus conferred through the instrumentality of Signor Sella was that of the order of SS. Maurice and Lazarus—a distinction offered, not only in recognition of the scientific attainments of the Local Director of the Geological Survey of Great Britain, but also as a mark of the appreciation of his services to Italian officers sent at various times to England on missions of scientific inquiry.

The end of this month of Alpine rambling concluded Ramsay's journeys abroad as an active geologist. For eight years he did not again leave this country. He had now practically accomplished the foreign travel of his life, and though he was able in later years to revisit some of the scenes which he had traversed in the full vigour of manhood, it was rather with a view to rest and change, or, where any scientific work was attempted, it was more for the purpose of testing conclusions already made than with the view of fresh exploration and new deduction.

CHAPTER IX

THE period of Ramsay's life on the history of which
we now enter embraces a space of about ten years.
During that interval he was mainly occupied in the
duties of the Geological Survey, finding time and
ability for fewer extra-official labours than he had
been able to accomplish before. His routine work
was not relieved and enlivened by the inspiration of
Swiss mountaineering; but he continued to perform
it with faithful persistence, and to superintend his
staff with the same firm and friendly hand.

It is one of the duties of the President of the
Geological Society at the end of each of the two
years of his tenure of the office to read an address,
which may either deal with the general progress of
geology during the previous twelve months, or may
treat of some special branch of the subject to which
the writer has particularly given his attention. For
some years past Ramsay had been brooding upon
what Darwin had so well enforced—the imperfection
of the geological record. He was struck by the
extraordinary gaps in the succession of organic remains, even where there was no marked physical
interruption of the continuity of sedimentation. And

he connected these gaps with geographical changes of which no other trace had survived. He had made a communication on this subject to the American Association at the Montreal meeting, which had attracted considerable attention among those present. He had afterwards made it the subject of cne of his evening lectures to working men at Jermyn Street. But no full exposition of his views had yet been made public. He therefore chose ' Breaks in the Succession of the British Strata ' as the thesis to be worked out in his two successive presidential addresses, taking the Palæozoic systems in the first year (1863), and the Secondary and Tertiary systems in the second (1864). Some account of these essays will be given in the concluding chapter of this volume.

In the months of January and February 1863 Ramsay gave a course of six evening lectures to working men in the Jermyn Street Museum on the Physical Geology and Geography of Great Britain. These lectures were taken down at the time in shorthand, and were shortly afterwards printed and published as a small volume. Unfortunately, the lecturer's state of health at the time prevented him from correcting the proofs with adequate care. The book consequently appeared full of inaccuracies. But the nucleus of a valuable handbook was there, and in later years its author was able to revise and enlarge it, and it now forms his well-known and admirable treatise on the *Physical Geology and Geography of Great Britain.* Even in the original tract the geological reader can perceive the outlines of many deductions regarding the growth of the surface topography of the land, which the author was able subsequently to work out more fully. The publication of this book marks a distinct epoch in its writer's

scientific career.   Thenceforward, while he continued
to take interest in all geological problems, and more
particularly in those which were engaging the attention
of his colleagues in the mapping of the Geological
Survey, it was the origin of scenery which had for
him the supreme attraction.   The history of lakes,
river-basins and valleys, the influence of geological
structure on landscape, and the effects of that structure
and of its accompanying topographical contours upon
the people of the country—these were the themes
which now engaged his thoughts, and on which he
loved to speak and write.

The old elasticity of mind which in the past had
enabled him to get through so much mental as well
as bodily work still refused to return, and though in
congenial society he could once again be the liveliest
and brightest of a party, he was apt to suffer from such
weariness as made even the simplest duties irksome.
Writing to me on the 5th May 1863 from Dolaucothi,
whither he had gone for a little rest, he says : ' I had
begun to consider recovery doubtful, but I now think
"there's life in the old dog yet."   All the while I could
eat, laugh, sing, fish, and walk a little (three or six
miles), but still I had misgivings.   Oh the charm of
this country and its pleasant friends !   Since breakfast I have been at a magistrates' meeting, seeing two
affiliation cases disposed of, and then engineering a
brook with the young ladies.   This country is full of
drift, with scratched stones and erratics going up to
600, 800, or 1000 feet, maybe higher.   But I have
seen no clear section of it, and do not know if it is
stratified.   I considered it so long ago, but I would
like to confirm it.'

The improvement in his condition was not maintained during the summer, and he looked forward with dismay to the winter, when the necessity for lecturing would once more meet him face to face. As his lectures were not written out, but were delivered merely from notes, which he changed and brought up to date from year to year, he always felt that the success of a lecture depended almost entirely on his condition at the time when he had to speak. Even up to the end, though the subject was quite familiar to him, and he could have discoursed for hours about it to a group of friends, the formal lecture to a miscellaneous audience, and still more to a company of students, was a severe mental strain to him. When it was over he would come out of the lecture-room sometimes so weary that he could only go home and rest. The prospect of the winter session of the School of Mines was, therefore, at this time so dark to him that he seriously proposed to resign his lectureship, if that could be done without pecuniary loss. He felt that if relieved from all teaching duty, he could devote himself with more undivided energy to his duties in the Survey, and that the change would be better for the Survey as well as for himself. 'If the Treasury throw out my proposal,' he wrote to me, 'then I am where I was; and as I do not intend to die, I suppose I must put on half-steam. I wish they could be, consistently with official etiquette, a little more liberal in the matter, for it is hard to begin to go back when one has served twenty two years and more, and is half a century old, especially when one's *Survey* work has been well trebled.' After some months of suspense, 'the everlasting No' of the Treasury was duly received. 'So there it is,' he wrote again, 'and I suppose when February comes I shall try [lecturing]. I feel, I am glad to say, even better than when you saw me last, and it may, perhaps, not be too much for me.

The field-work of the Survey was now in full march through the remaining tracts of the southern counties of England, and Ramsay took an active interest in it, and in the fascinating problems of physical geography which it elucidated. On the 7th November he wrote to me : ' The deevil a holiday have I had since I saw you. I have been I don't know where, but at Wellingborough of late, and Sittingbourne and Tunbridge. On Monday I go with Hughes and Whitaker to look at and arrange about Tertiary mapping between Folkestone and Dover, and then to Lindfield to see the last of the Weald, that is to say, of the solid rocks there. . . . By the way, I think I have given up the marine denudation of the Weald. Atmosphere, rain, and rivers must ha' done it. I'm coming to that, I fear and hope, and hoping, fearing, trembling, regretfully triumphant, and tearfully joyous with the alloy of despair at my heart, and the balm of a truthful Gilead spread upon the struggling soul, bursting the bonds of antique prejudice, I yet expect to moor the tempest-tossed bark of Theory in the calm moral downs of Assurance.'

The second presidential address to the Geological Society was read on the 19th February 1864. At the Anniversary this year the Wollaston medal was bestowed on Sir Roderick Murchison for his great services to the science of geology, and it fell to Ramsay's lot as President to present it. Briefly and gracefully he summed up the work of his chief, and added a little personal touch that gave a special charm to the incident. ' Perhaps on this occasion,' he said, ' I may be pardoned for recalling the memory of a time I well

RODERICK IMPEY MURCHISON

remember, when of all the geologists of weight, you, Sir, were the first who held out the hand of fellowship to me, a young man, when four-and-twenty years ago I was struggling to enter into the ranks of geologists.'

With the close of his second Anniversary address the reign of the President of the Society came to an end. Ramsay vacated the office, and was now relieved of duties which, though not onerous, impose sometimes considerable strain on the occupant, and consume not a little of his time.

His views on the origin of lakes involved him in controversy which at this time he was little fitted to wage. Murchison, in his presidential address to the Geographical Society, had vigorously opposed the glacial theory of lakes. Ramsay had refrained from replying to other criticisms, feeling that if his views were correct they would prevail, and that if they were not, no amount of partisanship on his part would save them from dissolution. But when his own chief put out 'an exceedingly authoritative protest' against his theory, he felt that it would almost be uncourteous on his part to remain silent. Accordingly, he wrote a temperate but cogently-argued reply, which appeared in the October number of the *Philosophical Magazine.* His letters about this time are full of reference to the subject, showing that though he published little, he was following with the most lively interest what was said on the subject by others.

He wrote to me on the 15th May: 'Altogether I am quite pleased with the rapid progress the lake-theory has made. Lyell amazes me in the matter. He told me the other day that it must be wrong, and he believed that the hollows were due only to the disturbance of the rocks. . . . Have you brooded patiently

for six months without ceasing over that passage at
the end of Jukes's memoir on the Irish rivers, in which
he discusses the valley of the Rhône above the Lake
of Geneva? It is admirable and true, and by'r lakins!
he never saw the location! Tell me not where is fancy
bred, but after my Frankland change of climate article
comes out,[1] if any other good sound argument occurs
to you that I have not used. Bauerman has drawn
a wheel so true that Best has to put a heavy weight
on it to keep it from running away!'

BEAUMARIS, 30*th July* 1864.

MY DEAR GEIKIE—I am as busy as man can be,
and am really getting fast on with that big Memoir,
which I trust will be for fifty years a text-book to
the Silurian geology of North Wales. I have read
Sir R.'s counterblast in proof [above referred to], and
I told him I must reply to it. How on earth can he pit
Dawson against Logan? Does he remember also that
'he always thought' that Switzerland was another case
of water-drifting? For his protest and Lyell's I care
not a rush. Lyell for years scarcely believed Agassiz,
and used to have a special anti-Darwin chapter till
after the great book [*Origin of Species*] came out. He
is afraid of time now, and none of them know anything about denudation and the true physical behaviour
of rock-masses. I lately had a very satisfactory letter
from Hooker on the subject. The worst of it is that
one can scarcely hope to convince them, or the old
geological world generally. You can't make a colour-
blind man see colours. None of them ever mapped a
country, as we have done, and disturbed countries to

[1] Dr. Frankland's paper on the ' Physical Cause of the Glacial Epoch' will be
found in *Phil. Mag.* May 1864.

them will still owe their mountain features to disturbance alone.—Ever sincerely,   ANDW. C. RAMSAY.

In the sixth edition of his *Elements of Geology*,
published in January 1865, Lyell noticed the theory of
the glacial origin of lake-basins, and adduced various
arguments against it. Ramsay once more broke
through his resolve not to get into controversy, and
replied to these arguments in a paper contributed to
The Philosophical Magazine for the following April.
These controversies among the geologists were
cleverly indicated in good-humoured caricature by an
artist in *Punch*, who portrayed some leading characteristic of each combatant. Murchison sits in front
cross-legged throwing up three globes like an Indian
juggler. Lyell to a rapt audience of hammers illus-
trates the origin of terrestrial features by breaking
open a globe and lifting up a large fragment of it.
Ramsay, on the other hand, is busy by himself in a
corner sitting astride his globe, and digging out his
valleys and basins with a big spade.[1]
But though these disputes seem to bulk large in the
scientific work of the day, they really occupied a very
subordinate place, and certainly in Ramsay's daily
work they were not allowed to take up much time or
thought. While he remained in London, the editorial
supervision of maps, sections, and memoirs left him
but little time for extraneous work. His health being
now rather better, he could once more push on the
completion of the bulky Memoir on North Wales.
His part had been finished, but the palæontological
appendix by J. W. Salter was still incomplete. That
able but uncertain and procrastinating naturalist had

[1] *Punch*, 23rd September 1865.

resigned his appointment in the Survey during the summer of 1863, and it was difficult thereafter to secure his continuous services for the completion of his part of the Memoir. But at last, towards the end of 1865, Ramsay could write and date his preface, and the work was finally issued to the public early in 1866. It was the most detailed piece of writing which the Geological Survey had yet published, and it contained deductions and speculations of the greatest interest in theoretical geology.

The work of the Royal Commission on Coal, of which Ramsay was an active member, demanded a great deal of time during the five years from 1866 to 1870. Besides the numerous meetings of the Commission and of its committees, he undertook much additional labour in preparing, with the help of the staff of the Survey, maps, sections, and other data for the use of the Commissioners. Now and then, however, some less technical application of geology would arise to enliven the routine work of the office, as when Dean Stanley asked whether the geologist could throw any light on the history of the Coronation Stone at Westminster, round which so many old legends hang. Ramsay wrote to me about this request as follows : 'Yesterday I was at Westminster Abbey with the Dean, specially to examine the Coronation Stone from Scone. It is a reddish-grey sandstone, with three pebbles in it, one quartz and two dark ones of a doubtful substance, which may be Lydian stone. It is a hewn stone, with chisel-marks on it, and looks like a stone originally prepared for building purposes. Macculloch says it was taken from Dunstaffnage to Scone by Kenneth II. I see according to your map Dunstaffnage stands on Old Red Sandstone. What is

its colour and character there ? Macculloch says the
stone is calcareous, and so it is. I am going to write
a short report for the Dean, so please let me know
soon.'

On the 2nd April 1866 the Royal Society of Edinburgh awarded to Ramsay the Neill prize 'for his
various works and memoirs published during the last
five years, in which he had applied the large experience acquired by him in the direction of the arduous
work of the Geological Survey of Great Britain to the
elucidation of important questions bearing on geological science. The presentation was made by the
venerable President, Sir David Brewster, and Ramsay
attended in person to receive it The ceremony was
fixed to take place at the same time as the visit of
Thomas Carlyle to Edinburgh as Rector of the
University, when he delivered his memorable address,
and when the degree of LL.D. was conferred on three
distinguished teachers of the Jermyn Street School—
Tyndall, Ramsay, and Huxley. One of the features
of this visit, which Ramsay remembered with special
pleasure, was the dinner of the Royal Society Club.
The Royal Society of Edinburgh, like its sister societies
in other parts of the United Kingdom, has its dining
club, limited in number of members, who comprise
the leading resident fellows. But the distinguishing
feature of the northern fraternity is that, while it per-
mits few toasts and no speeches, its proceedings are
always enlivened with songs, often written for the
occasion. For many years it has boasted a succession
of song-writers, one or two of them gifted with great
humour, some of whose songs are known far and wide
beyond the limits of the Club. The post-prandial
efforts of Lord Neaves, unmelodious but infinitely

witty, belong to a rapidly-vanishing past, but Sir Douglas Maclagan remains to delight his privileged listeners. His 'Battle of Glen Tilt' will be popular in Scotland as long as cultured conviviality holds a place in the country. Ramsay heard that and other famous ditties, and used to speak enthusiastically of the way in which the philosophers of the north play their 'high jinks.'

There was another gratifying presentation a fortnight later. The staff of the Survey gave their esteemed Local Director a handsome gold watch as a mark of their appreciation of his long and devoted exertions in the cause of the Survey, and of his personal kindness and helpfulness to themselves.

At the meeting of the British Association at Nottingham in 1866 Ramsay again led the geologists as President of Section C. Since his previous tenure of the office, ten years before, a custom had crept in that the presidents opened the business of the sections with a specially composed address. He had been called unexpectedly and rather late in the day to occupy the chair, and had not had time to prepare such an address as he could have wished to deliver to his brother geologists. He therefore discoursed to them generally upon the influence of geological structure on external topography, and more particularly upon the influence of igneous rocks. He introduced, but with some hesitation, his views of the origin of some so-called igneous rocks, such as granite, from the action of heat, 'with the aid of alkaline waters.' He also found a place for his doctrine regarding breaks in succession of life, and proclaimed himself once more a thorough uniformitarian.

After the meeting he sent me the following account of it :—

My dear Geikie—I had a week in Anglesey after the British Association meeting, and yesterday brought up wife and bairns. I shall stay for a Coal Commission meeting on the 11th [Sept.], and if nothing come of that to interfere, shall immediately take the field thereafter. The British Association meeting was a good one, and I stayed at Newstead Abbey, and slept in the poet's bedroom !

> In the poet's bed I slept,
>   And out o' the bed i' the morn,
> Out o' the bed I crept,
>   And blew my sounding horn ;
> Then down the turret stair
>   I winded in my glory,
> And light winds raised my hair
>   As I entered the refectōry.
> And oh for the muffins and tea,
>   Beef, ham, and venison pasty,
> The jam and the honey o' bee,
>   The marmalade so tasty !
> And ever at dinner again,
>   I swear by heaven and hades,
> We quaffed the bright champagne,
>   And jabbered with the ladies ;
> And the lights shone overhead,
>   And the coats of mail they glinted
> On the wall o' the hall where we fed,
>   Nor meat nor liquor stinted.
> No more have I to say,
>   Though the words could come by milliards,
> I presided in C all day,
>   And all night I played at billiards.

Yours ever more,
ANDW. C. RAMSAY.

The general tenor of his life among his colleagues in the field during the years up to the end of 1871 can best be gathered from his letters, from which a few selections are here given :—

KING'S ARMS, SHEFFIELD,
13th August 1865.

MY DEAR GEIKIE — . . . I have been all about
the universe; at Rowsley with Dakyns; there and at
Hathersage and the Snake Inn with Green. That
country beats cock-fighting, for it has no drift, is 2000
feet high, and otherwise ought to have ice, and has
none. Then I have been at Todmorden, the deadly-
lively, with Hull. I made two speeches at Manchester, and have also been at Kirkby and Dent and
Kendal, and am now here with Tiddeman.

See the last *Reader*, yesterday's, and expire. At
least Sir R. will, when he sees what I, being in my
right mind, have bequeathed him in my last will and
testament. When I leave this I want to see my wife
and babbies a little. They have been in Anglesey a
month, and I have not seen them for considerably
longer. . . . I sent my review of Campbell ['Frost
and Fire'] to Edmonston and Douglas [for *North
British Review*].

THE CITY OF THE DEAD, VEL DURHAM,
31st October 1866.

MY DEAREST WIFE — Luckily it rained to-day
when we got to this City of Silence, and therefore,
instead of starting for the hills, I had time to see it,
which I have been doing for three hours and a half,
and yet have left a deal unseen. You can concentrate
your energies on the architecture, for there are no
people for certain to look at. Here and there a
ghostly figure comes out of a corner and as
suddenly disappears, but whether these shapes are
human mortals' or not, I am unable to guess. To
wind up with, we have just come from church, where

certainly we did hear some sort of angelic melody.
But oh ! the grandeur of the Cathedral, all Norman
from end to end, excepting a sort of Lady Chapel of
very early English on the east, and, what is more, the
whole is almost unaltered Norman.   Three towers
hath it, one grand central one, and two at the west
end, which take away your breath with a sense of
beauty.   The great interior columns are marvellous to
behold, and the roof is grandly groined.   The vast
pile overlooks the river, and the west front extends far
down the bank, so that a wonderful dignity of height
is given to the building.   Then the bishop's palace
(now, alas ! a seedy college)—a vast pile, castle and
palace in one, partly Norman, and the cloisters, the
close, and lots of other things, which I must see
another day when I can make the acquaintance of
some local antiquary, if such there be in Durham.

DUNFORD BRIDGE, SHEFFIELD,
27th November 1866.

MY DEAREST WIFE—This is a bad place to write
from.   The reason is, that the post comes in at break-
fast-time, and in these short days we are in a great
hurry to get out, and when we come home again
across the moors the post has gone.   After dinner no
human being writes letters if he can help it.   The
above gives the reason why I did not write yesterday,
and may be the reason why I will not write to-
morrow.   But to-day I have received several letters
so important that I must stay in a couple of hours to
answer them. . . . The letters of most importance
were from Sir R. and Best.   The Duke [of Bucking-
ham] and My Lords are making sweeping changes, to
which I must reconcile myself, and I believe I can do

U

it without grumbling, and possibly even with tolerable satisfaction.

First, Scotland is to be raised to a special branch like Ireland, and Geikie is to be Director. Second, I am to get another £100 a year, to continue in charge of England and Wales, to drop the 'Local' before Director, and to be ranked as Senior Director. I am to have two 'District Surveyors' under me, who will be Aveline and Bristow; two first-class senior geologists, eight second-class, and the rest as before, except that we shall have a large addition to the staff. There would be no use objecting to anything, even if there be anything to object to, for the Duke and My Lords have ruled it, I believe, without reference to Sir Roderick. . . . I think I have written Sir Roderick a very good letter, without any grumblings at all. I have only compared myself to the Emperor of Austria, losing not Venice, but his German native dominions, and increasing his revenue thereby, and I have approved of all the other details.

HAZELHEAD, SHEFFIELD,
4th December 1866.

MY DEAREST WIFE — This is an awful day of wind and rain, and this is my tenth, and I hope my last letter.

From a very official letter Sir R. wrote me, I was afraid he had taken amiss the way I took these changes. But to-day I have had a very long and pleasant letter from him telling me that that was by no means the case, and that he wrote the short official simply because the subject was strictly of that nature, and he was communicating a copy of Cole's official bearing My Lords' pleasure. He also tells me that the importance of my position is very much raised, seeing that I shall have three times as many men to command as Jukes, and four times as many as Geikie. To this I reply, not satirically, that I feel the compliment of being considered able to do four times the work of other people, and hope it will be duly considered when pension time arrives. . . . The gale is tremendous, and the rivers are flooded.

The changes in the organisation of the Geological Survey referred to in these letters were the most important that had been made since the foundation of the service under De la Beche. The staff in Great Britain was divided into two, Scotland being made a distinct branch of the Survey under a separate Director. The title of Local Director for Great Britain being abolished, Ramsay became Senior Director for England and Wales. Jukes remained as before Director for Ireland, and the corresponding office in Scotland was given to the present writer. A new grade, that of District Surveyor, was created, in order that separate areas in which a number of the staff were at work might be more continuously supervised. The number of assistant geologists and geologists was largely increased, and it was arranged that there should be one geologist for every three assistants. When the new appointments were all filled up the Senior Director had under him a staff of thirty-seven men, the Director for Scotland nine, and the Director for Ireland fourteen.

As a consequence of this transformation, Ramsay ceased to have any control over the progress of the work in Scotland, and no longer paid his annual visit of inspection to the surveyors north of the Tweed.

But the number of men whom he now had to superintend in England was larger than he had ever had before. On reflection, he strongly disapproved of the increase in the staff, and he particularly condemned the way in which it was planned and carried out.   Though his long experience gave him a special claim to be consulted in any important changes in the organisation of the Survey, he never heard anything definite as to what was in contemplation until the whole scheme was matured and adopted.   He used to speak bitterly of the difficulty of procuring the authorised number of new men, for he felt sure that a good geological sur veyor could not be manufactured by a board of professors, nor even by a crammer, and could not be discovered by any ordinary form of examination.   The recruit, properly equipped by his education, could only acquire his fitness for duty by practical training, and it was, in Ramsay's judgment, impossible with his force of old hands, constituted as it was, to train at once half their number of new men.   He would have preferred adding to the force by degrees, as good men could be found and educated for their duties.

It must be acknowledged, however, that the Department of Science and Art in proposing, and the Treasury in sanctioning, this great rearrangement and augmentation of the staff of the Geological Survey, were sincerely desirous to further the objects for which the Survey was instituted.   They wished that, with as little delay as possible, the public should be put in possession of a general geological map of the whole country, and this end could not be attained for many years unless the force were largely increased. There was an additional reason that had much weight with the Lord President of the Council.   For some

years the Geological Survey had been carefully distinguishing and mapping the various superficial
deposits which, in the earlier days of the work, it had
not been thought necessary to discriminate. Apart
from their great scientific interest, maps of the surface
geology had innumerable advantages of a practical kind.
They gave information as to the nature and distribution of soils, and were thus of value for agricultural
purposes. They were of essential service in the construction of reservoirs, and generally in questions of
water-supply. They were of great utility in the laying
out of roads and railways; and they could be made
to furnish valuable evidence in relation to drainage
and sanitary matters. The importance of such maps
being recognised by Government, it was desired to
afford greater facilities for their production. It was
now arranged that the practice of mapping the superficial deposits simultaneously with the solid rocks
underneath them, which had been introduced into the
Survey some years previously, should be continued
over all the unsurveyed districts, and that, as soon as
surveyors could be detached for the purpose, the
tracts already surveyed where the surface-formations
had not been separated should be re-traversed for the
purpose of inserting them. By this means a general
agronomical map of the whole country would be provided, which would be of much service for farming
purposes, land-valuation, drainage, water-supply, and
many other practical affairs of life. These designs
have since that time been steadily kept in view, and a
large part of the country has now been completed.

   The changes in the Survey staff could not come
into operation until the beginning of the financial year,
that is, the 1st April 1867. Steps had been taken

before that time to obtain young men who gave promise of becoming efficient surveyors.   But, as Ramsay
had contended, it was extremely difficult to procure
the required number at once, and some time had
passed before he could announce that his comple-
ment was complete, and a still longer time before
he was able to replace the incompetent new-comers
and make his corps efficient.   There was much dis-
agreeable detail to be attended to before all these pre-
liminaries were settled, and his letters show that it
gave him a good deal of vexation.   But his gaiety of
spirit made even these worries sometimes a subject of
merriment.   His letters to myself were at this time
more frequent than usual.   A few of them are inserted
here :—

LUNNUN, *5th February* 1867.

MY DEAR BELL-THE-CAT[1]—We must have a profound talk over the colouring of Ayrshire, for there
will be plenty of fault-finders ; and as it belongs to my
reign (old Saturn), and as my aged eyes may never
see the Empyrean (Auchendrane[2]) again, we must settle
it among us, while yet, like the Centurion, I may say
to James (the Caledonian apostle), Come, and he
cometh.   Let him come, then, with all his maps, and
that will do for the blooming Peach's as well, and all
will be settled before the Jovial times begin.   If need
be, Bone in a day will draw in the lines (in pleasant
places) on a clean copy, and we will decide and colour
the rest.

---

[1] It will have been seen how playfully Ramsay used to vary the names of
his colleagues when he wrote to them.   The Christian name of his correspondent
on this occasion suggested the well-known sobriquet that was given to the great
Earl of Angus at the end of the fifteenth century.
[2] See the reference *ante*, p. 248.

Poor Jukes is in a sort of semi-despair about all
this business, and considering that he will be adding
ten more Irishmen to his already Irish lot, I don't
wonder at it.   His chief man lately informed him that
he had given up taking and recording dips, as he
found it to be useless!   Jukes simply longs for the
day when he will be able to retire, from age, and wear
out the fag-end of his days, unworried by Irishmen
and Boilermen, and I considerably sympathise with
him. . . . Oh for an hour of brave old De la Beche,
in his best days, to look ahead and provide for the
future!—Ever sincerely,   ANDW. C. RAMSAY.

15th June 1867.

MY DEAR GEIKIE—I had barely time to write you
yesterday about your summons by Sir R.   Jukes is
exceedingly fidgety.   He has not a man in Ireland
that he can trust to training others.   Also, they are
all so unruly, that without rules (printed) every one
will be in rebellion.   Even on this side of the water
I have no doubt we are all frightfully mismanaging
everything without knowing it.   At least, I have no
doubt I am, and I see no reason why you should not
be doing the same.   If you feel conscious that you are
not doing the same, that merely proves that you are
so blinded by ignorance and cockyness that you don't
know when you are doing mischief.   At least, I believe
that is the case with me.   Therefore everything must
be reduced to printed rules.

Now I am of this way of feeling, viz. I don't
want to have any duties, and I don't want to do them ;
and if it so happen that you are of the same opinion,
then it may fall out that Jukes may get printed rules
for Ireland, and leave us to that ancient unwritten

law which is the Lion of the North, and the bulwark of the Hammerers' faith.

I think I have now expressed myself in clear scientific language, and therefore you will dine with us on Thursday at half-past six.—Yours ever sincerely,

ANDW. C. RAMSAY.

KIRKBY LONSDALE, *3rd July* 1867.

MY DEAR GEIKIE—Yours received. Papers glanced at, but not yet fairly read.

The rule here is out at 9 A.M. ; no letters written before breakfast, except in cases of fire, murder, rape, and robbery. Home to dinner, and the post just going (as now), and too lazy to write after dinner, except in cases of abduction, stabbing, perjury, and earthquakes.

To-day we have been in a river, the Greta, from ten till five. When too deep for skipping and missing the stones skipped at, Tiddeman carried us across on his back (Hughes and me), because Tiddeman wears knickerbockers. I understood these villain Carboniferous rocks (Upper, Middle, and Lower Coal-measures ; Gannister beds and Millstone grit) better than I ever did before, and so did all of us. When you don't see a rock for miles except in a river, and that river is generally full to the brim and more, then there is usually Tartarus and Thomas to pay, without coin in your pocket. To make sure to-day, we all plunged into a pool to see what was in the bottom, but as we never got there, heaven only knows whether it is shale or Millstone grit.

If I get to the Railway Hotel, Newcastle-on-Tyne, by Saturday (which is on the cards), then I'll spend part of Sunday reading your brief.—Ever sincerely,

ANDW. C. RAMSAY.

To Miss Johnes he writes on the 17th July from Wirksworth : 'Since I left London twenty-four days ago, I have been staying at Kendal, Kirkby Lonsdale, Newcastle, Belper, and here. Kendal is a woollen-making place, but one charming day we spent on Windermere and in the neighbouring valleys. Kirkby Lonsdale is charming beyond expression. It lies on the River Lune, which is more beautiful than Alph, the sacred river. There is no trade in the town, and the people are very good people, parsons and all ; the gentry are hospitable round about, if you give them a chance, and the inn is old-fashioned, full of daughters, lively yet sedate, who, with their very handsome old mother, do not leave their guests to the mercy of servants. I sometimes think of taking Louisa there some day on our way to Scotland, that she may know what an English river is like.'

The British Association met at Dundee in 1867, and was attended by a large concourse of geologists. Ramsay formed one of the number, and though he read no paper, he took part in the discussions and excursions. He was especially pleased to revisit St. Andrews, where nine years before he had worked for three months at the Welsh Memoir, and where he had made many acquaintances. His old friend, Robert Chambers, who had come to live in the antique university town, was present at the banquet given by the Senatus to the excursionists, and afterwards had a reception at his house. This was probably the last time that Chambers and Ramsay met each other. Chambers looked already much broken in health, though he kept still his interest in geological progress. He died four years afterwards.

In the spring of 1868, in the intervals of examining

candidates and lecturing, Ramsay took the occasion of
the publication of a new edition (the tenth) of Lyell's
Principles of Geology to criticise that work in two
articles in the *Saturday Review*. Resuming the quotations from his letters, we may note that on the 18th
March he wrote to Mrs. Ramsay : Your father would be
about as busy as I am if he had to preach six sermons
a week, had, besides, twenty-four curates to superintend—six with him and eighteen constantly writing
letters—two of them rebellious, with also a bishop
staying in his house constantly consulting with him,
besides having about four magistrates' meetings a week
to attend. These last are my Coal Commissions and
Councils.'

LONDON, 15*th May* 1868.

MY DEAR GEIKIE—Your argument about recent
disturbances *in re* lakes is a good addition. I have
long given up taking any notice of those who oppose
me. They are impenetrable, and I feel so sure I
am right, that I can well afford to leave the rest to
time. But many people have a pernicious fashion of
stating that De Mortillet and I came to the same conclusion the same year. I wish somebody would some
day contradict that for me. He says that the lake-
basins existed *before* the glacial period, but how formed
he does not say. They were then filled with gravel,
etc., and the glaciers scooped out that—a very different
sort of story, and one that in no way grapples with the
subject. Did you see my two reviews of Lyell's
first volume of the *Principles* in the *Saturday* of the
11th and 18th April ?—Ever sincerely,

A. C. RAMSAY.

Ten days later he wrote to me further regarding

the opposition to his lake theory : ' All the objections make no impression on me, and I feel it best to leave them alone as far as I am concerned. But I still hope and intend to apply the view to *all time*—past, present, and future—and a good deal beyond at both ends.

' You will see a lot of curious papers in the volume which I will send to-morrow. I stayed at Bonn two months. I have given Zirkel of Bonn a letter of introduction to you.[1] He is going to the Western Isles. He is a fine young fellow, and a Professor at Lemberg ; he would like, too, to see some work. . . . You must take old Hibbert on the Eifel if you go there. Van Dechen's big map of the Drachenfels region is not very good ; there is an explanation of it in German. Also, the Government geological maps of all the Prussian Rhine region are published. I can lend you some. Be sure you see the Miocene coal at Brill, half-way between Bonn and Cologne. I'll give you letters if you like. Go and see the Moselle and its tributaries—the best case of valleys cut in a table-land that I know. You must march through the Eifel— 6s. per day, living like *les coques qui se combattent.*'

LEEDS, 18*th September* 1868.

MY DEAR GEIKIE—Late, late, so late ; but I will venture now to reply to yours of the 4th, which work, laziness, and sometimes imperfect health, and consequent demi-semi-depression, prevented my sooner replying to. Not that I am ill, and yet I am just something or other. It may be that it is only Age with creeping claw that has claught me in his clutch.

[1] See *ante*, p. 262.

If so, so much the worse for Age, for he has got hold of a bad lot.

Like the men of the '45, I have been 'out' since the 29th June, all but a fortnight, which I spent in Anglesey; and also, like the same men of '45, I have had a controversy with the king, not King Cole, but King Roderick of Siluria. . . . Some people wonder why I did not reply to Sir R.'s last in the *Geol. Magazine* about denudation and lakes, but I think it is better not to 'condescend upon' it, as we Scotch lawyers say. But why should he be always troubling our Israel? Is he afeard that we are becoming rebellious satraps?

I did not go to Norwich [British Association Meeting]. I stayed away a-purpose to keep out of any excitement. Last year did me no good, and giving evidence at the end of June for four days before the Coal Commission for four hours and a half per day, together with an immediate march and long hours in the country during the hottest weather, have not improved me. So I stayed away from Norwich. D—— writes me that the *advanced scientific thinkers* did themselves and science no good at Norwich. How, I have not heard; but I can well believe it of some of our friends. . . .

I know the Strahleck, having been over it, and very steep it is on the descent from the Col down to the surface of the glacier on the Grindelwald side. But it is very different in different years. Hinchliff slid down on the snow from top to bottom. I think it took us an hour to go down on the rocks. . . .

We have done a deal of work hereaway, and are fast moving up northwards in a broad line, in the hope of forming a union with the Northumberland and

Westmoreland men, before you can say whew. We have begun in the Vale of Eden, and will by and by invade your dominions, if you don't mind your eye.— Ever sincerely,   A. C. RAMSAY.

BLANCHLAND, 24*th September* 1868.

MY DEAREST WIFE—I write to tell you that I am living in a fragment of an ancient abbey, placed on the banks of the Derwent, far up the stream. The house is now an inn, and our window looks out on a plot of grass that may have been in the middle of the cloisters. The modern church, a fragment of the old one, *re-muddled*, looks on our grass; and pear-trees, trained against the walls, the fruit of which the monks ate, writhe their old branches all about the stones. Such relics of a beautiful antiquity always fill me with a sort of regretful feeling. If it had only been possible to preserve them! How many lovely spots there are in England that one never heard of till one gets in among them. Howell came with me from Hexham ; we drove over the hills, twelve miles, after four o'clock yesterday. At Hexham there are also the remains of a grand abbey. The transept and the chancel are entire, and are used (though abused), but the nave is gone. It is as big as many a cathedral, and noble Early English in style.

I must tell you a story of our friend Noumeran, the Japanese. He had a post-office order sent to the country, and when he signed his name the postmaster insisted that it would not do. 'You must sign your Christian name as well.' 'But,' said Noumeran, 'I am not a Christian; I am a Pagan.' Amazement of the postmaster, who only knew of Pagans before as of dragons, or griffins, or fabulous monsters of some sort.

Howell told me a story of Disraeli. Vernon Harcourt asked a Conservative friend, 'How can you and your party follow such a man?' 'We look on him as a professional bowler,' was the reply.

The men wait.—Your most affectionate,

ANDW. C. RAMSAY.

DENT, KENDAL, *4th October* 1868.

MY DEAREST WIFE—I begin another letter to you to-night to tell you something about this place; it is so beautiful. The valley is five or six miles long, 'well watered.' While below it is full of lovely green meadows, bordered with trees and dotted with old white-washed houses of the dalesmen, all around great bare hills rise to heights of 2000 and 2300 feet. And the little town is so quaint, irregular, and clean, with its village church and absence of shops, that all combined fill the mind with a sense of repose and old-fashionedness, but rarely met with now in toil-worn England. And the people are so nice. Last night we spent with the Sedgwicks in the house where old Adam was born. Mrs. Sedgwick is very pretty, and only about your age. She has at home six girls and a little boy. They all crowd round Hughes, and climb on his knees all at once.

I have written to old Adam Sedgwick telling him how pleased I am to be in his old home, and how kind Mrs. Sedgwick is, and I hope he will be pleased with my letter.

This vale of Dent filled Ramsay with delight, which breaks out again and again in his letters. Thus to Miss Johnes, on the 11th October 1868, he writes: 'Dent is not on the outskirts, but in the core

of the world, and the farther you recede from it in
concentric circles, the nearer you get to the outposts of
" civilisation falsely so called." Dent town and the
valley of Dent make a kind of paradise to a man
troubled with cares of Geological Surveys and Coal
Commissions.   Fancy a valley some six or eight miles
long, well watered, with green sloping pastures and
noble trees, with great peaceful, solemn hills all
around ; noises unknown from the outer world, no
sounds, in fact, but those made by winds and running
rivers, or dropping rains and cattle, and the voices of
" the kindly race of men," and church-bells o' Sundays.
All the children are clean (very) ; all the men are
stalwart and frank, honest and brave ; and all the
women that are not beautiful are comely, some of
them stalwart too.   Men, women, and children, Danes
by descent, are fair, with blue open eyes—" statesmen," the men part, in the northern sense of the term
—frank and respectful, for self-respect makes folk
respectful to others.

'I have been away from home for four and a
half months, as human mortals usually count them,
but to me the time looks like four and a half mortal
years, and I long to see Louisa and my children
again.'

His journeys of inspection now ranged over the
whole breadth of the northern counties of England.
On the 21st November 1868 he wrote to me from
Barnsley : 'Since I saw you I have been at Newcastle,
Bellingham, Morpeth, Ponteland, Richmond, Harro-
gate, Pateley Bridge, Otley, Bolton Bridge, Skipton,
and here.   I have seen, besides geology, Ripon
Cathedral, Knaresboro, Kirk Hamerton (real Danish
or Anglo-Saxon), Bolton Abbey and Fountains Abbey,

besides Skipton in Craven, where, as you very well know, "there's never a haven." '

These inspection tours brought him into the midst of delightful scenery, interesting geology, varied historical associations, and pleasant society—a combination of attractions that never failed to show him at his best. Professor Hughes, who now holds the Woodwardian Chair at Cambridge, was then one of the staff with whom the Director had many excursions, and who has kindly supplied me with the following recollections of his chief. Speaking of the evenings after the day's tramp was over, he says : ' Ramsay always threw himself heartily into whatever game or amusement of any kind was going on, and thus got an insight into the life of his men, and helped to make things pleasant for them with their neighbours. So agreeable a companion at a dinner - party, and so considerate and obliging a guest at an hotel, was always welcome, and every one asked when he was coming back, and tried to arrange little plans to make his stay pleasant. He loved a game of cards or billiards, which he played to win, not with the bored expression of one who did it just because he was asked to, or merely to kill time. He was very fond of chorus-singing, taking the bass with a good deal of skill and great earnestness. Even when there was no entertainment going on he was generally very lively all the evening.'

LONDON, 26*th October* 1870.

MY DEAR GEIKIE—I have been away since July, and only came home on Monday last. I have had an awful battering on the Yorkshire hills of late in thunder, lightning, and in rain (Williams). . . . I am very well, and have been, barring an eye,

which is now rather better than it was before it got
worse.

Wife and babbies all well, and rejoicing more over
the desperate willain who has returned than over
ninety-and-nine just men who stay in the field and do
their work. I'm off again on Monday for Lancashire,
about Preston, etc. etc., and shall be thereabouts for
two or three weeks; after that to Grantham and the
Oolites. I won't be much at home before the end of
December. Sir R. looks well.

Oh the dales, the dales, the Yorkshire dales!
Lovely, luvely, loovely! Cock-fighting be hanged!
The terrassic system! The Carb. Limestone is a
myth proper, and the Yoredale Rocks ditto. They pass
into limestones in the most unprincipled manner, and
now the limestone runs up to the Millstone Grit, and
now, to use a strong expression, it doesn't. Lithology
is the only science, and as for definite horizons, they
no more exist than nadir or zenith, the equator or
Fergus the First.—Ever sincerely,   A. C. RAMSAY.

JERMYN STREET, 16th *February* 1871.

MY DEAR MISTER G —I have gotten yourn. I
believe I have Rütimeyer's book, and that I looked it
over, but I am a poor ignorant son of a sea-cook, and
cannot read German. But I get bits translated for
me by Ella or the Missus. As Rütimeyer does not
agree with me, of course he is wrong! Desor's paper
(French) I have read, on the origin of Jura valleys,
lakes, etc. etc., and thought it 'Walker.' I do not
think I have seen any German treatise of his on the
subject.

I am glad you are writing that paper for
Nature; it will come in rather pattish. Prestwich

x

sent me his remarks to read—what he is going to say when he hands over to me the Wollaston medal, and he says nowt about the lakes. They must be still too strong for his geological stomach. But he has swallowed other things handsomely, and remarks that in the matter of Palæozoic ice I long stood alone. He may live to swallow all the 4000 feet of Swiss ice that scooped out lakes, and also all the big northern ice-sheet that buried two-thirds of the northern continents.

Do you think Rütimeyer shows good cause for his dislocations in the Alps without good mapping done?

My paper on the Old Red, etc. etc., has not yet been read. I suppose it will come on upon the 22nd March or thereabouts. They print the papers now entire for convenience before they are read. It does not follow, I believe, that they will necessarily be printed in the Journal.

Now I must go to prepare a lecture for 2 P.M. I gave one last Monday night on the Origin of the River Systems of England, and the audience liked it better than I did.

If I write it,
And I like it,
I will send it
To the Royal.
If they like it,
And they print it,
I will send it
To my Geikie.
If he read it,

And he likes it,
I will like it
All the better;
For my Geikie
Is judgematick,
And he knoweth
All that differs
B's between
And feet of bullocks.

—Yours ever,   A. C. RAMSAY.

About this period Ramsay's pen was more than usually busy. The problems suggested by red strati-fied deposits like the New Red Sandstone and Old Red Sandstone had often been considered by him, and he discussed the subject in two papers communicated in January and March 1871 to the Geological Society. One of these dealt with the red rocks of Palæozoic age, and the other with those of later date. He was likewise turning his thoughts more frequently and earnestly to the history of topography, and especially to the origin of river-valleys. He gave a series of lectures on that subject during this year, and afterwards condensed the substance of one or two of them into a paper on the 'River-courses of England and Wales,' which was read before the Geological Society on the 7th February 1872.

Much anxiety was felt during the year 1871 as to the health of the distinguished Director-General of the Geological Survey On the 30th November 1870 he had a stroke of paralysis, which at the time deprived him of the use of his left side. But he rallied so far as to be able to take carriage exercise, and to attend to a good deal of business. It was evident, however, that he would never again be fit to resume his place in the scientific world, though he might possibly linger long. Trenham Reeks, his faithful secretary, and Registrar of the School of Mines, used to see him at his house daily, bring official and other letters, arrange about the answering of them, and despatch frequent bulletins to members of the staff as to the condition of the chief. Murchison never again set foot in the Museum in Jermyn Street.

But as there was no immediate prospect of serious change in Sir Roderick's condition, Ramsay took the field among his men in the spring of 1871. Some further letters from him show what he was doing and thinking about during the summer and autumn of this year.

24*th March* 1871.

MY DEAR GEIKIE—I write a second note.   If you refer to my book on North Wales, you will see that I state that the Lingula Flags and Cambrian are conformable, and pass into each other, and that the Llandeilo and Bala beds lie unconformably on both. *Officially* I still call the Lingula Flags Lower Silurian, because of the Director-General's classification, but theoretically I consider the Lingula Flags more closely allied to the Cambrian.   In the first paper I sent you you will see, however, that I consider the Cambrian (below Lingula Flags) as a fresh-water formation.   The Llandeilos and Balas are, however, nearly as closely connected with the Tremadoc Slate and Lingula Flags as the Upper Silurian is with Llandeilo and Bala beds.   The Tremadoc Slate I consider an upper part of the Lingula Flags.—Ever sincerely,   A. C. RAMSAY.

KING'S ARMS, KENDAL,
1*st October* 1871.

MY DEAR GEIKIE— I'm smoking a pipe on a Sunday.   Hallelujah, hallelujee! I send you my two last [papers on Red Rocks] to Edinburgh, not knowing where you may be.   I have had a very pleasant letter from De Koninck about them.   He will write again, but at present he seems equally surprised and pleased. Besides twelve sent here and there in England, I think I will devote the rest of my copies to Continentals and Americans, for Englishmen can read them in the Journal. . . .

I came here last Tuesday, and, weather permitting, have been daily among the Silurian Green Slates and Porphyries.   The more I see of them, the more am I convinced that all of them I have seen form part of a great purely subaerial volcano = the Welsh marine or semi-marine set. I am assisting at a bit of mapping which Aveline cannot make up his mind about, and I have made up mine. . . . I hope to leave this on Tuesday or Wednesday for Kirkby Stephen. It is tough work here, driving twelve miles and then climbing, like a climbing boy, mountains from 2000 to 3000 feet high during equinoxious gales. Sir R. continues just the same.—Ever sincerely,

A. C. RAMSAY.

LONDON, 13*th December* 1871.

MY DEAR GEIKIE—Austen has decided to take the Presidency of the Geological Society.[1] That is well. But the Society is not flourishing in papers. I am glad we are to have a strong President, and we must try to get a strong council, made of men of mark. When Dallas wrote me the other day, asking if I had no papers to fill the void, I replied that I had none ready, and of course could not write for the sake of writing. I also said that they might have more were it not that authors of theoretical papers were afraid to send them in for fear of the fatherly care of the Council. Green's last paper was squashed, by —— in particular. You will see it in the Geological Magazine. I have a great mind to send in a paper entitled ' The Wonderful, the Councillor,' with illustrations, by Rutley, of living examples. When at Clapham (Ingleborough) last Friday I explored a cave 800 yards long with Tiddeman.—Ever sincerely,

A. C. RAMSAY.

---

[1] In the end he was obliged to decline the office.

LONDON, 16*th December* 1871.

MY DEAR GEIKIE—Yesterday I took a leaf out of the Book of Othello, and became perplexed in the extreme,' all along of a miserable trusteeship that I hold. The consequence was that all day I stood prostrate at the feet of Europe, having there*fore* an aversion to all legitimate business, till in the evening I plucked up ' hart o' grease ' and wrote to S. Kensington asking about the Survey Annual Reports. I have to-day had an answer saying that My Lords meet on Monday, and the question will come before them. 'That is all we know on Earth, and all we need to know.'

Having done this, I went at half - past four and played with Herbert Spencer a game of billiards at the Athenæum. He beat me. Being beaten, I went home and ate a turkey, and then proceeded to lay about me all round and make everybody miserable, all because a woman 200 and odd miles off is an ass, and gives me a deal of deilish bother.

The poet [his son Allan] came home yesterday from Uppingham with a prize for mathematics under his oxter. We are all very well at home, both the cats having disappeared. I hope they are not at the bottom of the water butt !

Several of the drift maps have been engraved. We are now in a state to publish. I will tell Bristow to send you a coloured specimen copy for your criticism before any are issued. I hope we may manage to do so early next week.—Ever sincerely,

ANDW. C. RAMSAY.

LONDON, 11*th January* 1872.

MY DEAR GEIKIE—Yesterday in the Council of the Geological Society I proposed Croll as a proper

man to receive the Wollaston Fund for the year. . . .
The President and others hailed my proposition.
One objection raised was that Croll's researches
involved no personal expense. Prestwich and I
thought that of no importance; but nevertheless if you
can tell me anything on that score I shall be doubly
armed—

> And on the top of opportunity,
> Quell the base scullion rogues, whose envy dull
> Would squash the light of Genius, and instead
> Display a dirty, spluttering, farthing dip,
> And swear that 'tis the sun.

So look alive, my pigeon, and help in this good
cause. . . . I can do nothing till my third edition of
Physical Geology and Geography is in the press. I
am now at the last lecture of it. I will turn it into
chapters. It will be nearly twice as long as it was,
and so much modified (I hope improved) that it may
almost be said to be a new book.—Ever sincerely,

A. C. RAMSAY.

Sir Roderick Murchison died on the 22nd October
1871, and the office of Director-General of the
Geological Survey once more became vacant. When
he accepted the appointment it was with the expressed
intention of holding it only for a short period. I
will tide you over a few years,' he said to Ramsay
at the outset. But he retained the position for sixteen years and a half. Of his geological labours, and
more especially of his connection with the Geological
Survey, a full account has been given elsewhere.[1]
These matters have therefore been only cursorily
referred to in the foregoing chapters.

[1] *Life of Sir R. I. Murchison,* 2 vols., 1875.

The death of their Director-General necessarily gave rise to considerable anxiety among the officers of the Survey. They hoped that their friend and colleague, who had been passed over at the time of De la Beche's death, would not be passed over again. Rumour, of course, was busy with reports of various kinds—the Jermyn Street establishment was to be broken up, there was to be no new Director-General, or an outsider who was variously named was to be once more put over the service. But, happily, these predictions proved to be false. After four months of suspense, Ramsay received from Lord Ripon, who was then Lord-President of the Council, a letter dated 26th February 1872, asking him if it would be agreeable to him that he should be nominated to the vacant post. It was explained that the delay had arisen because various questions connected with the several branches of the establishment in Jermyn Street had been under consideration. The appointment thus offered to him did not embrace the School of Mines. A great scheme was in contemplation for the formation of a College of Science at South Kensington, for which the Jermyn Street School would form an excellent nucleus ; and it was therefore considered expedient to sever the tie which from the beginning had united the School of Mines to the rest of the organisation planned and carried out by De la Beche. A fortnight later I had the following note from Ramsay : '16th *March.*—I was yesterday summoned to attend a meeting of the [Committee of] Privy Council at South Kensington. The result is I am Director - General of the Surveys, Museum, and Mining Record Office, Bristow succeeds me as Director for England and Wales, Howell succeeds Bristow (as District Surveyor),

and Ward gets up to be geologist.' Thus at last he
had attained what had been the height of his ambition.
After thirty-one years of faithful service he was now
placed officially at the head of the organisation of
which he had so long been the life and soul.

CHAPTER X

BEFORE we follow the subject of this biography in his new sphere of duty and responsibility it will be of advantage to look for a moment at the state of the Survey when he was called to be the head of it. He was no longer to be immediately responsible for the personal supervision of the field-work. It is interesting, therefore, to consider in what condition he left the mapping in England and Wales, and how far the Survey had advanced in Scotland and Ireland, which were now to be under his jurisdiction.

In England and Wales the only untouched tracts were Norfolk and Suffolk, with portions of Essex and Cambridgeshire, the greater part of Lincolnshire and Yorkshire, and the north-western portion of Cumberland and Northumberland. The field-work was being pushed forward in the six northern counties, Northumberland, Durham, Cumberland, Westmoreland, Yorkshire, and Lancashire. A group of surveyors was busy in Lincolnshire, Nottinghamshire, and Leicestershire, and another in Essex, Hertfordshire, and parts of the home-counties. As for some years past all the surface-geology had been traced upon the maps as well as the outcrops of the older works underneath, a considerable part of England had now been surveyed for Drift.

In Scotland the survey, extending westward from the area where Ramsay began in 1854, had stretched across the island from the mouth of the Firth of Tay to the mouth of the Clyde, and southwards to a wavy line drawn from Berwick-on-Tweed to Wigtown. There were still large tracts of the counties of Berwick, Roxburgh, Dumfries, Kirkcudbright, Wigtown, and Ayr to be surveyed. The Highlands had not yet been touched.

In Ireland all the country south of a line drawn from Clew Bay to Dundalk had been surveyed, and most of it had been published. The ground to the east of a line from Dundalk to Lough Neagh had likewise been in great part surveyed and published. All the northern part of the country was still unmapped, including the north-western tracts of Mayo, Donegal, Londonderry, Tyrone, nearly the whole of Antrim and Fermanagh, with large tracts of Sligo, Leitrim, Monaghan, and Armagh.

The duties of Director-General of the Geological Survey and Director of the Museum of Practical Geology are necessarily to a large extent administrative. But as far as possible he keeps himself in touch with the field-work by personally visiting the districts that are being mapped, and becoming acquainted with the details. By making himself familiar with the problems encountered in each of the three kingdoms, he is enabled to bring the experience of one branch of the Survey to bear upon the difficulties of the others, and thus to ensure more rapid progress as well as more harmonious results over the whole United Kingdom. All maps, sections, and memoirs, are submitted to him before publication, and in this way a good deal of editorial work is imposed upon him. The amount of

Survey correspondence is thus necessarily large and constant.   The Director-General is, further, the official channel of communication with his own and with other departments of Government, as well as with the general public.   Endless are the applications he receives for information or advice on geological questions.   At one moment he is asked for assistance in supplying an arsenal or fort with water; at another he is requested to inform a government board where a prison or a workhouse had best be placed. One Colonial government inquires of him whether in his opinion water is likely to be obtained at a particular spot which he may never have seen or heard of. Another sends home a box of earth and stones with a request to know whether the material affords hopeful indications of gold.   A landed proprietor in England asks him why no coal has been found on his estate, another forwards a parcel of 'specimens,' and wishes to know what useful minerals he may look for in the places from which they were taken.   A third sends a so-called 'fossil,' dug up on the estate, in the belief that it is some unique treasure, when it proves to be merely a lump of inorganic concretion.   In numberless questions of drainage, road-making, railway-engineering, water-supply, choosing sites for buildings, and other matters where a knowledge of geology has a practical bearing, applications are continually made to the Geological Survey for assistance.   It may readily be believed that the Director-General is thus involved in a large amount of extraneous business, besides that which more properly arises from his ordinary official duties.

The quantity of letter-writing which now fell upon Ramsay, whether by his own hand or by that of a secretary, was often so large that it left him hardly time for other avocations. Amid the multitude of letters he was glad to keep them as brief as might be. He could comparatively seldom indulge in the pleasant gossiping epistles which hitherto he had been wont to send to his friends and colleagues, but restricted himself more and more to the absolutely essential business. Nevertheless for a few years he contrived to find opportunity for putting on paper some of the observations he had made on the origin of the features of landscapes, and for communicating papers on this subject to the Geological Society.

Except that he 'was more involved in official routine, and had less time for inspection of field-work and for original research of his own, his life as Director-General passed much in the same way as that of Director had done. One year slipped away like another, only marked by longer or shorter spells of London life. But he had one great resource in a house at Beaumaris left to Mrs. Ramsay by an old friend. To this retreat he betook himself more and more with his wife and family. There, away from the thousand distractions of town, he attended to his correspondence and worked at the geological or literary undertakings which duty or choice imposed on him. The mountains of Caernarvonshire rose in front of his windows to remind him, as he felt himself no longer young, that he had been a good climber in his day, and that on their flanks and among their Cwms and crags he had done some of the best geological labour of his life.

Ramsay had now gained the position which for so many years he had wished to reach. But it must be confessed that the reward came to him too late to enable him to profit by it as he would have done had it been conferred ten or fifteen years sooner. He had probably never quite recovered from the effects of that disastrous break-down in 1860.   Had he been able to free himself from the burden of his lectureship at the School of Mines, he might perhaps have been restored to complete health, and have escaped from that mental weariness which his friends and colleagues used sorrowfully to watch as it increased upon him during each succeeding session of the school.   Even the advancement to be Director-General did not throw off this incubus.   The income of the appointment was reduced by the amount of his salary as professor, and he was compelled to go on lecturing for five years longer, until the Treasury at length agreed to restore the emoluments of the office to what they had formerly been, and to permit him to resign his lectureship.

In the early autumn the new Director-General made an official tour in Ireland, in order to become personally acquainted with the various officers in that part of the United Kingdom, and also to see some of the more salient features of Irish geology, of which as yet he had not been able to obtain any knowledge by actual examination in the field.   The following letters supply us with some pictures of the tour :—

LARNE, *29th September* 1872.

MY DEAR MRS. COOKMAN—I am at Larne, in Antrim, some twenty miles or so north of Belfast, and have with me my good friend John F. Campbell, ycleped of Islay.[1]   And this is how it happened.   Being in London against my will, *in re* a fight in the Irish branch of the Survey, I heard that Campbell was also

---

[1] John F. Campbell, born 1821, died 1885, author of *Tales of the West Highlands*, was also fond of geological observation.   He had travelled far and wide, and was the author of the picturesque and entertaining work *Frost and Fire*, besides other volumes of travel.

in London in a dismantled state. . . . So here we are!
At Dublin I transacted a deal of business, saw some
marvellous antiquities at the Irish Academy, and then
with my colleague Mr. Hull, Director for Ireland, we
started for Dundalk, where besides seeing the geology,
we visited Cuchullin's Rath and saw the grave of Fin
M'Coul; worked our way to Newry; from Newry to
Warrenpoint, and joined there a fine young fellow
called Traill, one of the Survey. He looks something
like what I did when I joined the Survey, only he
is much handsomer, sings a great deal better, but
cannot jump so high. We saw the Carlingford country
and all the Mourne Mountains, and progressed to
Newcastle, staying two or three days at each place.
We were out on the mountains or on the sea-cliffs every
day, and have been battered by equinoxial gales from
every point of the compass. . . . I do not expect to
be home for a fortnight or three weeks, for by easy
stages I have to continue this royal progress till
we get to Galway, and then back across the great
central plain of Ireland to Dublin.

The Irish trouble to which I alluded has lost me
three weeks, otherwise I proposed going to Germany,
and perhaps taking Ella with me. But I begin now
to see that it will be too late for this year. I have
such a pretty problem in my mind, if I could only
tackle it.

FLORENCE COURT, *5th October* 1872.

MY DEAR GEIKIE—I got your letter last night on
our arrival here, Hull and self. I have been making
a grand round with Hull for more than a fortnight. . . .
Here at Lord Enniskillen's we stay till Monday. I
have learned a great deal about Irish geology and
physical geography previously to me unknown. John

Campbell stuck by us as far as Armagh, when he branched or shunted off to speak Gaelic to the folk in the north-west.

Thanks for your remarks on Arran. I have seen the Mourne Mountain and Slieve Croob granites, and one of them took away my breath. [Here follows a rough sketch showing an upper cake of contorted and baked Silurian strata with basalt dykes, underlain and cut off by a mass of granite.] That is in the Mourne Mountains, and the drawing represents the crest of a large mountain a mile or two long ; it is one of many such.

When I saw it drawn in a section I could scarce believe it. But I saw it afterwards on the ground, and it is true.—Ever sincerely,   A. C. RAMSAY.

The loughs of Ireland, as might have been expected, roused the enthusiasm of one who had studied lakes so closely, and had been involved in so much controversy about them. Regarding those of Fermanagh he tells Mrs. Cookman : ' These lakes here (Upper and Lower Lough Erne) first took away my breath, then made my hair stand on end, and then confused my intellect so lamentably that I doubt if I will ever write sense any more. They are the most curious lakes I ever saw.'

DUBLIN, 11*th October* 1872.

MY DEAR GEIKIE—Since I wrote to you I have been at Sligo and Boyle, and I now write to say that near Boyle I saw Old Red Sandstone, which doubtless is Lower, and it contains bands of felspathic lavas precisely like those of the Pentlands and Oban. I have seen no Old Red Sandstone that is not Lower and the Carboniferous lies highly unconformably on it.

I now also know a deal about the great Carboniferous Limestone plains of the middle of Ireland, and something of the coal-fields. Ireland must have been somewhat like Finland long ago, before so many of its lakes got turned into peat-mosses. I have also partly realised the Shannon and a lot of other odds and ends in a three-and-a-half weeks' tour among Hull's men. I have seen all the staff but two, and a very nice set of fellows they are. I leave to-morrow night, and get home on Tuesday, I hope.—Ever sincerely,

A. C. RAMSAY.

One of the periodical tasks of the Director-General is to receive the reports of the field operations, of the indoor work, and of the Museum for the year, and to prepare from them his Annual Report of progress, which is sent to the Department of Science and Art to be presented to Parliament, and published in the annual blue book of the Department. At the end of the year these various returns are prepared by the officers of each branch of the Survey and the curators of the collections in Jermyn Street, Edinburgh, and Dublin, and the first duty of the chief after the beginning of January is to master their contents, to procure additional information or correction where needed, and to work the whole into a narrative of all that has been done during the previous twelve months by the different establishments under his control. Buried in the pages of a blue book, these Annual Reports are much less widely known than the labour spent upon them entitles them to be. It was now Ramsay's turn in the early part of 1873 to compile the yearly statement. His personal familiarity with the men in the field and their work enabled him to attack the most difficult part of the task with spirit and success. In these labours of routine, and indeed in all the official work of his office, he received constant loyal and efficient aid from Mr. Edward Best, who, originally appointed as an assistant geologist for service in the field, had been transferred early in his career to the office in Jermyn Street, where he acted as general secretary in charge of the correspondence and the issue of publications. Mr. Best's long experience made him familiar with all the details of the history and progress of the Survey. He was a general favourite among the staff, and for many years served as the right hand of his chief.[1]

Ramsay still occasionally contributed an article to the *Saturday Review*, and gave a Friday evening discourse at the Royal Institution. In the pages of his favourite 'weekly' he wrote pleasantly about the history of Great Britain, taking a much wider view into the past as well as into the future of the subject than the ordinary historian is content with, and adding a caution to our statesmen for the benefit of their successors fifty thousand years hence, when a new glacial period shall begin to banish man from the northern half of Europe. ' It behoves the Minister for the Colonies,' he concludes, ' to see that our inter-tropical possessions are kept in good order for the coming migration, for the fortunes of the British Islands will then be far below zero. One cold comfort remains — the universal northern ice-sheet may possibly solve the Irish difficulty.'

---

[1] He joined the service in 1855 under De la Beche, and retired from it on 31st March 1893, carrying with him the affectionate regard of all his colleagues.

Another occupation of the same winter was the writing of an article on the River Po for *Macmillan's Magazine*. As his mind dwelt so much now on rivers and their operations, he was led to recall what he had himself seen of the workings of the Po and its tributaries on the southern flanks of the Alps, and over the vast plains of Lombardy. He likewise read extensively the literature of that great river. The paper which he now wrote was translated by Gastaldi into Italian for the Bulletin of the Italian Alpine Club, and separate copies of the translation were printed for the Italian Government, that they might be distributed widely among local authorities and others in Lombardy. In sending a copy of the Italian version of the paper to Mrs. Cookman, Ramsay told her that he had 'heard from Italy that the article has helped to stir up the authorities there *in re* their duty to their rivers and the people.'

Before the Royal Institution he discoursed on ' Old Continents,' and sent me the following account of the evening. ' I lectured last Friday [12th February 1873] on " Old Continents " to a very full house. As I treated it, the subject was quite new to every one there, and by good luck I was in the right humour for lecturing. I restricted myself to the great continental epoch between the close of the Upper Silurian and the end of the New Red Marl, and put all episodes in consecutive order. The act of lecturing on it suggested some new ideas which I did not broach, for I had quite enough to do without them in an hour. However, perhaps they may bear fruit in a paper for the Geological.'

Having been chosen by Murchison as his literary executor, and charged with the writing of his biography, I had applied to Ramsay for any of Murchison's letters which he could supply, and also for information as to the best way of procuring materials from some of the old chief's correspondents. He answered as follows: 'I have no influence with Sedgwick. We are very good friends, but he never quite forgets the Survey having turned his Cambrian into Lower Silurian, so aiding Sir Roderick, without specially meaning it. . . . I doubt if Hughes will be able to help you in that matter. Sedgwick is still sore about it. . . . I never saw Wollaston, but Greenough, Buckland, Warburton, and Fitton I knew. There ought also to be De la Beche, Sedgwick, old John Taylor, Whewell, Mantell, Major Clark, old Stokes, Sir Philip Egerton, Lord Enniskillen, Babbage, and others. They used all to have a jollification at Lord E.'s rooms in Jermyn Street after the meetings. Lord E. told me a lot of things last autumn, which I now nearly forget.'

Of the voluminous memoir on the Geology of North Wales, published in 1866, a new edition was now required, and its author set about the necessary preparation. The house at Beaumaris came then to be of more practical value to him than ever, for while it allowed him to escape conveniently from London, and to keep his family around him, it provided him with a home near the ground which he might have to re-examine. This new edition continued to be one of his main employments during the rest of his official life.

While at Beaumaris, in the summer of 1873, he made a short excursion to St. David's, the geology of which had in recent years been brought into prominence by Mr. Salter and Dr. Hicks, whose con-

JOHN W. SALTER

clusions did not quite coincide with those expressed
on the maps of the Geological Survey. Writing to
Mrs. Ramsay from that remote cathedral town on the 3rd
August, he says : ' To-day (Sunday) we have been at
the cathedral, and I sat in my old stall and sang bass.
But the music has sadly fallen off. The organ is dismantled because of the repairs of the church, and there
is only a harmonium, and the singers are diminished.
Scott is slowly restoring the building, but there is still
a great deal to do, with as yet insufficient money.

   ' To-morrow we take a boat and coast along for
eight or ten miles to re-examine the coast section.
The weather is splendid, and it will be delightful.
We have first-rate boatmen, one being the captain of
the lifeboat. . . . Now that I am here, it would never
do to leave the country without bringing the geology
of St. David's (which is now exciting so much attention) up to the modern mark. Considering how
ignorant I was in 1841, I wonder I did it so well.'

   Three days later, writing to the same correspondent, he tells her : ' Probably we will start to-
morrow, drive up to Fishguard, and thence to Cardigan. I shall refresh my memory on geological
points by the way. . . . This is a moist, hot climate,
like Cornwall. Your very clothes get damp, and your
gummed envelopes get also damp and seal themselves.'

   On his return to Beaumaris he sent me the following account of the trip into Pembrokeshire : ' I
have been for a fortnight at St. David's seeing all
Hicks's discoveries among the Cambrian rocks and his
Menevian strata, which form a grey band, 550 feet
thick, between the uppermost purple Cambrian grits and
the bottom of the Lingula Flags. Fossils numerous,

all of the same kind as those in the Cambrian beds, only some additional genera and species, but none or few common to the Lingula Flags. Etheridge I took with me, and David Homfray came also from Portmadoc. As I knew before, there are boulder-beds at St. David's, but I did not know that they contain chalk-flints, which are also found in Ramsey Island. The country is undoubtedly *moutonné*, and I saw on the coast in three places striations running N.N.W. and S.S.E., pointing, in fact, to the north of Ireland.

' My *Contemporary Review* paper, as regards substance, is in all important points my two Red Rock papers in the Geological Journal, only the subject is put consecutively. . . . I am satisfied that in Scotland there are two or three glacial episodes in what is commonly called Old Red. I have no doubt, however, that you will work it out, and I see no reason against a Carboniferous glacial episode. The day will come when all folk will allow a Silurian one too, which I long ago inferred from the rocks on Carrick Moore's land, and troubled his mind by printing the idea.

' Hicks I think is wrong about his Laurentian axis at St. David's. I believe the area is Cambrian metamorphosed into a kind of syenite, and that the granites there are, like those of Anglesey, also metamorphic. But I could not be supposed to see all that in 1841 when I surveyed the area.' [1]

He was able this summer to carry out at last his intention of visiting the Rhine Valley, for the purpose of studying the problem of its origin. Taking with him his eldest daughter, and accompanied by his sister and his nephew (now Professor William Ramsay of University College, London), he ascended the river

[1] See p. 172 and *note*.

from Cologne, and remained a week at Bingen, making excursions up the valley of the Nahe and the Rhine. Thence the party went to Strasburg, and Ramsay took some geological expeditions into the Vosges valleys for the solution of the question he had come to study. Being so near, it was impossible to resist the pleasure of seeing some of his old friends and former haunts in Switzerland. So with his travelling companions, he made for Basle, Lucerne, and Meiringen, crossing the Scheideck to Grindelwald, where he was much interested in the diminution of the glaciers since he had previously seen them. By way of the Wengern Alp, Lauterbrunnen, and Interlaken, the party reached Berne, where they remained some days taking excursions with the venerable Studer. They then made for Bex, where Ramsay, with eyes now quickened to perceive the profound interest of river-courses, was delighted to have an opportunity of examining the valley of the Rhône where it bends sharply round at Martigny, and farther down where the river is filling up the upper end of the Lake of Geneva with sediment. Returning to Basle, they experienced much kindness from that delightful veteran of Swiss geology, Peter Merian, and from Professor Rütimeyer, and then made their way homeward by the east side of the Rhine, through Heidelberg to Cologne. It had been Ramsay's intention to descend the whole length of the grand old river down to Rotterdam, but on looking into the state of the finances of the party, he found that they had just money enough left to take them straight back to London, which they reached by way of Ostend.

On his return he sent me (26th September) a few jottings of his doings : ' I got home last night, having solved my problem in a very different way from what

I expected. It was curious to find all the supports to one's speculative views crumbling away one after another. So I began again quite dispassionately *in re* the Rhine, and the result is that I think I have done the gun trick, which is too long to write about. I have also learned a deal of other odds and ends.' To Mrs. Cookman he wrote : ' Ella and I had a delightful journey. I saw at least five of my old friends in Switzerland, two of them, alas! over eighty years of age, but I rejoice to say quite hale and hearty. But I missed old Escher von der Linth, who is no more. I learnt heaps of things, and will send you a memoir when it is written and printed, on the physical history of the valley of the Rhine. If you and Betha will come out with my wife and me, we'll explore the valley of the Rhône from Geneva downwards, and next year do the Danube from its sources in the Schwarzwald to its mouth, and write joint memoirs on these subjects, for I am rather crazy about rivers just at present, and it will be of great advantage to the world if Louisa and you will get crazy too.'

The results of this brief continental excursion were quickly brought before the world. On the 4th February 1874 Ramsay read an account of his observations to the Geological Society in a paper on the Physical History of the Valley of the Rhine, and on the 27th March he gave a Friday evening discourse on the subject to the Royal Institution. A reference to his views on this question will be made in the succeeding chapter.

Next summer, as was usual now, he spent some time at Beaumaris, making excursions thence to re-examine ground for the preparation of the Welsh Memoir. With the company and assistance of Mr. Hughes, formerly one of his staff, but who had now succeeded Sedgwick as Woodwardian Professor at Cambridge, and also with Mr. R. Etheridge, and Mr. D. Homfray of Portmadoc, he traversed a good deal of ground in the district of Cader Idris, Aran Mowddwy, and Portmadoc. Rain and wind buffeted the party a good deal, but the Director-General declared that 'every day hardened his old legs more and more, and by the end he cared little for the fatigue.' In an account of his doings (9th August) he wrote to me that 'the necessity for a second edition of my *North Wales* is now urgent, and I am seriously at work making out a new line of division, that between the base of the Llandeilos and the top of the Tremadocs, or, *en grand*, between the Lingula Flag series and the Llandeilos. I have *accurately* traced twenty miles of it, and have for the first time (yesterday) got perplexed. We have been at work for about eighteen days.'

Some further information is given in a letter to his brother William, written from Portmadoc during a subsequent excursion. (13th September 1874): 'I am busy revising a deal of country and realising all the discoveries that have turned up since Selwyn and I were here more than twenty-five years ago. It involves the tracing of one new geological line that no one suspected long ago, and which I surmised must exist ever since Sir Roderick and I were in the north of Scotland some fifteen years ago.'

The progress of his work and the nature of some of his engagements during the year 1875 are told in the following letters :—

30th January 1875.

MY DEAR GEIKIE—I highly approve of your vindication of De la Beche [in proof-sheets of the *Life of*

Murchison[1]]. In fact, the one-sidedness lay all on the other side. . . . Last Wednesday Ward read a second paper at the Geological Society on the Cumbrian Lake-basins, going the whole hog *in re* their glacier origin. Bonney allowed that he could go that length, but that the theory in no way applied to the Swiss and Italian lakes. Since then I have received a letter from Gastaldi enclosing a MS. of a paper to be read at the Academy of Turin on Sunday, the 7th February, in which he proves that all the great Italian lakes were produced by glacier erosion, and giving good reasons for their special positions and relative sizes in relation to the valleys in which they lie, and also their relations to the Pliocene deposits of the valley of the Po. He wants my opinion on it before it is read, which I will send him on Monday. In the meanwhile, my wife is translating it, and I will publish it with his permission after he has read it. It is a very important paper.— Ever sincerely,   A. C. RAMSAY.

16th April 1875.

MY DEAR INFANTA [Miss Johnes]—I do not think you can have any idea of the good that my visit to Dolaucothi did to me physically and morally. I went on improving after I got home, too, extending even to the art of lecturing within the last few days. The pleasant quietness and the absence of all necessity for being agreeable in your house (!) did me a power of good. The dawdling by the quiet waters and among my old friends the individual trees, and the sitting opposite your father in the library, to say nothing about his two daughters all about the house—*ach, lieber Himmel!*—these things were good for a man.

---

[1] See vol. ii. p. 186, *note.*

You know that I was always soft enough in the head, but now I think I have got softened all over, not excepting the engine that drives the blood. I do not think I have even thought a cross thought since I came home, and I only hope I may always be able to keep up that decent frame of mind. You see I do not mind what I say to you or your sister. You are one of the Sisters of Charity, and I make my confession to you as I would to the Pope, honest man, if I happened to be intimate with him, and liked him sufficiently. I often think of that pleasant episode in my life which began when I first went to Pumpsaint, and has lasted up till to-day. That to me is a golden legend better than any that Caxton ever printed, for in spite of a few clouds, so much of it has been full of air, light, and sunshine. On the 26th July 1842 I first went to Pumpsaint, and there was no winter at all that year, nor for several years after. And even now there is no more of it than is perhaps good for one.

LONDON, *24th April* 1875.

My DEAR GEIKIE—Since receiving yours of 21st I have been very busy. . . . Last night I lectured at the Royal Institution on the Pre-Miocene Alps, and their subsequent waste and degradation, to a good audience. It is a difficult subject to make quite plain to a general audience, the figures are so large ; but though I was not quite satisfied with it myself, Sir Philip Egerton and others seemed to think I made it clear.—Ever sincerely,   ANDW. C. RAMSAY.

LONDON, *23rd July* 1875.

My DEAR MRS. COOKMAN— . . . As for me my life is rendered miserable by writing testimonials for men

trying for professorships in Australia, Japan, Aberystwith, and Eton—the latter a mastership. I have
had the nomination of a man to the University in the
capital of Japan, as Professor of Geology. Then I am
driven wild by invitations for self and lady to all sorts
of public soirées—three for this very night, to only
the quietest of which I will go. However, I have
made up my mind to dine with the Lord Mayor, and
I will take Louisa to the soirée of the Society of Arts.
. . . I have had a letter to-day from America, and the
'critter' encloses another sheet giving a sketch of my
life, and asking me to fill in some blanks and correct it
and send it back for publication in *Appleton's American
Cyclopedia!* I am laughing consumedly. — Ever
affectionately,   ANDW. C. RAMSAY.

During the summer and autumn he was again busy
in Wales, and from time to time he sent me tidings of
his doings. From his letters to me the following
sentences are taken.

'*27th July.*—There has been no practicable weather
till yesterday. It is fine now, and I start for Merionethshire on Thursday. The book is not in a state to
make any further progress till more work is done in
the field. I had brought it up to that point.

'*12th August.*—Here [Beaumaris] and in Mon-
mouthshire, Etheridge and I are hard at work. In the
mountains about Dolgelli I have Ward and Hebert,
having shown them the sections and the needful line
to add to the map. I am not fit for daily high
mountain work on a large scale now.

'The case stands thus: We have (1) Lingula
Flags, (2) Tremadoc Slates, (3) Arenig beds, which in
my Memoir are called *Lower* Llandeilo or Arenig beds,

above which in N. Wales comes the Caradoc series, of which the old-fashioned Llandeilos of Murchison may with propriety be considered a part. From the Tremadocs into the Arenig beds there pass about ten or eleven species; from the Arenigs into the ordinary Llandeilo and Caradoc beds eight species. I hear that you have equivalents of the Arenigs in the S. of Scotland, somewhere towards the Cambrian country. Can you give me any idea of their real relations to any overlying Silurian strata, and underlying strata, if any? . . . I have some evidence that the Arenigs of Caernarvonshire have overlapped all below, and lie direct on the Cambrians. It is in my *Geology of N. Wales*, but I think these Arenigs are there called Llandeilos.

' 14*th September.* — I scarcely think I have had enough of rest in the entire way. I am getting on with my new edition in spite of too much correspondence, and I have now got all the data except a scrap. The great Arenig and Lower Llandeilo line is done and run out to sea at both ends. I think I would almost rather write a new book than a new edition. Dovetailing is often so troublesome. I think (or hope) that I shall soon get to the last half, which may need but little alteration, except a few words here and there. As it turns out, a good deal of my book will be almost re-written. It will be a great improvement on the last edition. In the Welsh section, the trappy inter-stratifications are, of course, accidents, and sometimes, as at Criccieth, they are absent.

' 23*rd December.*—I am so head and ears at present in the River Dee (Wales) that I think I have got water on the brain. It is the most curious bit of physical history of any river I have yet tackled. It will make a chapter of my Survey Memoir on N. Wales; but I shall first send it to the Geological Society.   30*th*.— I have finished my Jolly Miller chapter [on the origin of the River Dee].   The results rather astound myself about the extremely early date of the river.   It has been running so long that the Rhine is a baby to it in age.

He had on hand at this time a number of geological memoirs bearing on his favourite topic of the origin of the superficial contours of the land.   The paper on ' How Anglesey became an Island' was read before the Geological Society on the 19th January 1876, and that upon the River Dee upon the 26th April following.   Then he was busy during the winter partly on the new edition of the Welsh Memoir, and partly on a revised and enlarged edition of his little volume on *The Physical Geology and Geography of Great Britain*.   He still also fired an occasional shot in defence of his theory of the glacial origin of lake-basins.   We resume the extracts from his correspondence.

On the 26th January 1876 he wrote to me : ' My Dee paper only went in [to the Geological Society] last Saturday, and therefore is not likely to be read for a month or two, in the time of the new President, Dr. Duncan, who assumes office on the 19th February. Austen would not take it, and there is no doubt that Duncan is the next best man.   Evans's selection of him gives general satisfaction.

' We are very deficient in volcanic rocks in England and Wales, excepting those of Lower Silurian date. These are all younger than the Tremadoc Slates, and begin in the Arenig series (Lower Llandeilo of Murchison).   If, as I have said, the red Cambrian rocks are mostly fresh-water or estuarine deposits, that may give some sign of an old valley; but in Wales there is a great thickness of Lingula Flags and Tremadoc Slates between the Cambrian strata and the volcanic series. . . . I know all the New Red series of England north of the Severn, and it contains no volcanic rocks. In Devon it does, see De La Beche's *Report on Devon and Cornwall,* chap. vii., where you will find something directly bearing on your question *in re* valleys and volcanoes.

'We have no Liassic or Oolitic volcanic rocks, and none in the Wealden, though that formation must have been deposited at the mouth of a great river-valley. Neither have we any Cretaceous or Eocene igneous rocks, though there is in these formations evidence of the mouth of another big river-valley.

'The Eifel volcanoes are, in general, on the top of a plateau, in which, however, there were pre-Miocene valleys, if these volcanoes be Miocene. . . . I think your letter gives me a glimmering of what you are thinking about in the matter of these old river-valleys and volcanoes, and the subject is quite a new idea to me, and will be to others when you work it out.'

To Miss Johnes he wrote on the 2nd March: 'I have pretty good news to tell of myself. This is my last year of delivering lectures in the Royal School of Mines. This will be a very considerable relief in point of work. As I had pay and fees as a professor, they cut off that part of the salary given to Sir Roderick, supposed to represent the Museum as distinct from the Survey. Now they are to add £300 a year to the Survey salary for the Museum, and cut off the Professorship salary, etc., and for that "crowning mercy" I am very glad, and so is Louisa. The fear is that by and by, having no occasion to lecture students, I may take to lecturing her instead.'

To Mrs. Cookman he reported that he had given his farewell lecture on the 9th May, 'fifty-five minutes being devoted to a broad resumé of geological subjects, and eight or ten to taking farewell of the students, and of my post of Professor of Geology. In that theatre I had lectured for quarter of a century, to say nothing of three previous years at University College. Glad as I am to stop, the severance cost me a sort of pang, and for a moment at the beginning of the valedictory, I almost thought I was going to break down—a tendency which my watchful wife observed, but which probably no one else did. In point of fact, I bit my under-lip, and swallowed something like a young potato in my throat. A sprinkling of extra strangers came, and all my children were there to hear Papa's last lecture.'

On the 22nd June he formally sent in his resignation of the lectureship, and the task which had once been so light and joyful, but which in these last years of failing power had become an increasing burden, was now happily removed. He told me that he would be financially a loser by the change, 'but the relief will counterbalance that.'

10th July 1876.

My dear Geikie—On Thursday I take my family to Beaumaris. Etheridge will join me by and by to have a bout at the rocks of Lleyn (north horn of Cardigan Bay), where I begin to believe I shall find the Arenig Slates lying directly on the Cambrian, without the intervention of the Menevian, Lingula, and Tremadoc beds, and involving a vast unconformity.

It is a most important point in British Silurian geology, as I have long attempted to show, and if I find this additional demonstration I shall be glad.

I shall be at work in Wales, writing the Memoir, with some field-work, for a good while, how long I know not. In September I go to Gibraltar for the Colonial Office and the authorities at Gibraltar, with an assistant. All expenses for both will be paid, and I asked nothing more, considering it a piece of duty. It is *in re* water. Now I would like much if you could spare your brother James to go with me. A few weeks will do it, for the surveying will be brief.

In accordance with his plan of work Ramsay this summer revisited many parts of the ground in North Wales which Selwyn and he had surveyed so long before, and with the assistance of some of his colleagues traced in the new boundaries. It is interesting to notice among his notes of these journeys that over tracts where formerly he had eyes only for the old rocks and their structure, he now looked out eagerly everywhere for the tracks of glaciers and ice-sheets, examined the rock-basins, and went carefully over the exposures of drift.

BEAUMARIS, 10*th August* 1876.

MY DEAR GEIKIE—We started at eight yesterday morning, and had twelve hours on the Caernarvonshire hills. We began by marching to the top of Moel Tryfaen to see the Cambrians and the shell-beds of Trimmer, 1150 [correctly about 1400] feet above the sea. These are undoubtedly true marine, beachy, false-bedded sands and gravels, overlaid by good boulder-clay, and have been much eroded before or during the deposition of the latter. It is a long story. We saw much moraine matter *en route* up. There is no description of that area in my *Old Glaciers of North Wales*, and no printed description of Moel Tryfaen gives an account of all that we saw. I shall write a note about it for the Geol. Soc., and also put it in my new edition of *North Wales*, when I have digested it. In the meanwhile, I see nothing in it adverse to my broader views in the *Old Glaciers of North Wales*.

From Moel Tryfaen we walked across moor and hill to Llanberis, that I might get a notion of the great extension of the slate-quarries since I first mapped the country some twenty-seven years ago. . . . I re-examined the Cambrian rocks at Bangor the other day, and found that the Arenig beds lie directly upon them, without the intervention of the Tremadoc Slates and Lingula Flags, as I have all along maintained. This is an important point for me *versus* mere stratigraphico-palæontological men who delight in finding errors in Survey views. I wish you could find Arenig beds directly on the Cambrians in the West Highlands.

On the 14th September he sailed for Gibraltar. With the assistance of his colleague, Mr. James Geikie, he made a careful survey of the Rock and a portion of the surrounding ground, steamed in a gunboat along the Spanish shores, crossed to the opposite mainland, and sailed for fifty miles along the African coast so as to get a bird's-eye view of the geology for purposes of comparison with that of the northern side, spent three days in Africa geologising and wandering among Moors, camels, and the picturesque but odorous streets and suburbs of a Moorish town. He was back in England on the 30th October.

On his return he set to work at once on the report
of his examination of Gibraltar with reference to the
question submitted to him.   But materials enough of
a more generally interesting geological character had
been collected which it seemed a pity to bury in the
pages of a departmental blue book.   The fellow-
travellers accordingly worked these materials up
into a conjoint paper, which in the spring of 1878 was
read before the Geological Society.

A pleasant incident diversified Ramsay's London
life during the winter of 1876-77.   His pupils at the
School of Mines had raised among themselves and
former students about £100, with which they purchased a set of three handsome silver dessert pieces,
half a dozen old Dutch parcel gilt spoons, and some
other table ornaments.   These they presented to their
much-esteemed teacher as an expression of their
gratitude and good-will for his eminent services to
the School, and for the benefit they had themselves
derived from his teaching and his influence.

How much his thoughts turned to foreign lands
and the geological questions there awaiting solution
is well shown in his correspondence at this time.
Thus to Mrs. Cookman he wrote: 'I have planned
a route to San Remo in the hope of going there to
fetch you home; viz. that Louisa and I first go to
Mulhouse or Basle, and then find our way down the
Saône to Lyons, where it joins the Rhône, then work
up the Rhône to Geneva, and back to Lyons, and so
down the remainder of the Rhône to Marseilles, with
a possible *divertissement* into Auvergne.   There is
something to find out about the valleys of the Saône
and Rhône that I know nothing about, and which I
think no one but myself has yet dreamed of.   On the

whole, I have always been a pretty good hand at scientific dreaming, and I believe this dream will come true, if I can only find time to work it out.'

A heartrending tragedy occurred during the autumn of 1876 in the family at Dolaucothi. The butler shot Mr. Johnes, severely wounded Mrs. Cookman, and afterwards committed suicide. As these were Ramsay's dearest friends, the event was a crushing trial for him, and in some measure saddened all his later life. His diary and his letters of this period afford touching proofs of the tender affection and deep sympathy of his nature.

But the vortex of London life swept him on. 'We are all well enough,' he wrote later in the year to his Dolaucothi friends, 'and, as usual, occupied with those innumerable busynesses which take up so much of people's lives in London, that anything like leisure becomes an unknown quantity. Of course, I am at work on a book in scraps of time, and if it were only finished I fancy I might breathe more freely, but I know that something else is sure to succeed it. The Survey men both of Scotland and Ireland are crying to me "Come," and go to both I must, some time this year.

'The invasion of scientific foreigners has also set in with unusual severity at an earlier season than usual. I invite them to dinner; some of them cannot come, and some do come, and then we have a Babel of languages. To add to that, we have got a German housemaid who as yet speaks no English.

'Since writing that last word, dispatches have arrived from Nova Scotia requiring immediate attention, the writer asking a letter from me, which, being shown, shall stimulate the Governor of Newfoundland to see the importance of certain work on the north-east coast of Labrador, whither my friend is being despatched from Newfoundland the dreary. He is the man who told Louisa that the worst dinner he ever had consisted of "cold eagle and badger-sauce."'

To the same friends, who were now on the Riviera, he writes : ' I grieve to say there is no chance of foreign travel for me this year. I must go to Ireland, and I must go to Scotland, and I have irons in the fire that must be got out and cooled, some of them, I hope, ere this year is much further advanced. . . . All of those valleys opening into the Alps, from the Dora Baltea to Como, have made a deep impression on me. I wish I could see them again, and specially with you two and Louisa. Besides the beauty, some curiously interesting points have come out since we were there. It is now known that the great lakes of Como, Maggiore, etc., were at one time fjords, like those of Norway. When the mighty old glaciers were busy scooping out *my* lake - hollows, the ends of them descended into the sea, and deposited their moraines there, for sea-shells are mingled with the material at the ends of the moraines. Then as the glaciers retired, the lakes became fjords, and I hear that, just as in the Swedish lakes, some marine species still inhabit the waters of Maggiore. There is a geological infliction for you! I give it you without remorse, for I know you to have a soul above buttons, unlike me, when once I wandered out round a lonely lake at the Grimsel in search of any kind of button, and found one of brass, by the margin on an ice-scratched rock.'

The journeys of inspection to Ireland and Scotland were duly made, and pictures of his progress may be gathered from his correspondence.

BEAUMARIS, 9*th September* 1877.

My DEAR GEIKIE—I got back on Thursday, after a month's stiffish work all about Ireland, from Kilkenny and Galway up to Portrush and the Giant's Causeway, and so south by Belfast back to Dublin. I have seen all the Miocene basalts in the north of Ireland, and from thence have had a view of Rathlin, Isla, Jura, the Mull of Cantyre, Ailsa, Arran, the Ayrshire mountains by Loch Doon, and a lot of small islands away north by Oban that I could not name. As for glaciation in Ireland, by Glendalough [Galway], etc. etc., good heavens!!! The sections at Glendalough and away north are the Silurian rocks of N.W. Sutherland, etc., quartzites, limestones, and all.

I shall be ready for the Cumberland and south of Scotland comparison with you and your men, Peach and Horne, and with Aveline, Ward, and Bristow.

The conference with his colleagues in the north-west of England and the south of Scotland was the last important conclave which Ramsay held in the field. The Directors for England and Scotland, each with two of their respective staffs, met him at Kendal, and the party journeyed through the more important geological tracts. Those of the number who had not been out in the field for some years with their chief saw with regret the marked failing in his vigour. The old brightness and kindliness were there as fully as of old, the merry laugh still rang out after the ready jest, and the lively talk, with interesting reminiscence and literary allusion, still charmed as they had always done; but the elastic step, the eager endurance, the sustained power of tracking the intricacies of geological structure had grown markedly feebler. I remember well the pang with which I realised as we climbed a hill-side above Derwent Water that my beloved friend, whom from my boyhood I had looked up to with pride and affection as the very embodiment of geological prowess, had now become an old man. He was then not more than sixty-three years of age, but a life of physical and mental toil and official worry had made him prematurely aged. At the end of the day when we got back to our inn he would often look exceedingly weary, and yet dinner would for the time revive him, and make him once more what he used always to be, the gayest member of a Survey gathering.

I remember that on the same occasion he showed how difficult it now was for him to keep pace with the onward developments of his own science. The introduction of the microscope as an adjunct to a field-geologist's equipment and the microscopic study of thin slices of rocks for petrographical determination had been recognised for some time by several members of his staff as absolutely essential for accurate mapping in regions of crystalline rocks. I had myself made use of the aid of the microscope for twelve years before this time, and J. C. Ward had adopted the same course in his study of the volcanic district of the Lakes. The party having dined, Ward and I had retired to another room that we might examine under the microscope some of his volcanic rocks, and compare them with the Palæozoic volcanic series of Scotland. We had been engaged on this task for an hour or two when Ramsay joined us. He sat rather impatiently watching us for a while, and then starting up, left the room after exclaiming, ' I cannot see of what use these slides can be to a field-man. I don't believe in looking at a mountain with a microscope.'

While on the journey through Scotland, he sent the following account of it to his friends at Dolaucothi : ' From one of the windows of this coffee-room [in Perth] I can see the Tay, full-flooded, rushing through the arches of that noble bridge, which reminds me of the bridge across the Moselle at Treves, only both river and bridge at Perth are more striking than those of Treves. Of a verity there is no denying the fact that the Tay is the finest river in Britain, with more water in it than even the Thames or the Severn, and such a varied landscape to flow through, with hills and cliffs, woods and swelling fields, all undulating and brae-like, except the noble haughs or meadows that here and there form the banks of the river, and of which the Inches of Perth (once islands) form such beautiful examples.

'I went to Keswick, Cockermouth, Carlisle, Hawick, Melrose, Galashiels, Peebles, Edinburgh, Leadhills, and Moffat, Dumfries, Kirkcudbright, Dumfries again and Edinburgh, and so to Arbroath, Stonehaven, Blairgowrie, and Perth. There is a catalogue for you, that beats Homer's catalogue of ships, or Milton's catalogue of devils. I hope ere a fortnight elapses to be in the bosom of my own family at Cromwell Crescent.'

Returning to London, he was soon once again in the midst of his ' new edition ' and other multifarious preparations. His paper on the geology of Gibraltar was read before the Geological Society on the 6th March 1878, and he gave a Friday evening discourse on the subject on the 24th May, which was his last appearance before the Royal Institution. The fifth edition of his *Physical Geology and Geography of Great Britain*, on which he had been engaged in intervals of leisure for several years past, was at length published in May of this year.

For some time previous to that at which this narrative has now arrived, Ramsay had suffered much trouble from an affection of the left eye, brought on in the first instance by overwork in lamplight, and aggravated by a severe wetting at the funeral of a brother-in-law. In the autumn of 1878 the ailment became so serious that, to save the other eye, it was necessary to remove the left one—an operation skilfully performed by his friend Mr. Whitaker Hulke, and borne by Ramsay with his characteristic quiet courage.

A month later he wrote to his friends at Dolaucothi: ' I am well enough to be doing much as usual, excepting that I do less work as yet. Then I have got a most lovely glass eye. You are not to quote Shakespeare, " Get thee glass eyes, and, like a scurvy politician, seem to see the things thou dost not." '

On the 16th June 1879 he received a gratifying telegram from Sella at Rome, that he had been elected a corresponding member of the Royal Academy of the Lincei, an honour which he specially prized, not only because of the famous Society which conferred it on him, but because it came as a mark of the kindly esteem of his friend, the illustrious statesman and geologist of Italy.

In the early summer of 1880 he succeeded in taking a brief holiday in a part of Europe which he had not before explored. Mrs. Ramsay and his second daughter had spent the winter at Hyères, and he met them on their way home at Aix les Bains. The traces left by the vast ancient glacier of the Rhône, which once overspread all that region, filled

him with astonishment and delight. He boated down
Lac Bourget, walked over hills strikingly ice-worn,
and picked up fragments of gneiss, granite, and other
rocks that had been brought down by the ice from the
heart of the distant Alps. He took the bearings of
the glacial striæ, observed the positions, sizes, and
composition of the erratic blocks, and saw so much as
to fill him with the strongest desire to return and
make a more complete examination of the district for
comparison with the old glaciated areas so familiar
to him at home.

From Aix the party made its way to Geneva,
spent a day or two there with the Swiss geologist, A.
Favre, and was back in England again by the 10th
June.

Ramsay had been elected President of the British
Association for this year, and the meeting was to be
held on the 25th August at Swansea. In the quiet
of his retreat at Beaumaris he prepared the presidential address. He chose a thoroughly geological
theme, and contrived to say a little on all the branches
of the science in which he himself had specially worked.
After a general historical introduction he launched into
the subject of metamorphism, and then into that of
the volcanic eruptions of former periods, whence he
naturally passed to the structure and relative ages of
mountain-chains. The salt-lakes of past times and
the recurrence of fresh-water conditions again and
again in geological history were next touched upon,
before the fascinating topic of glaciers and their
operations was reached. In summing up his discourse, the President professed once more his geological faith as an uncompromising Uniformitarian,
declaring that, from the period of the oldest known

rocks down to the present day, 'all the physical events in the history of the earth have varied neither in kind nor in intensity from those of which we now have experience.'

The discourse, though printed, was not read by the speaker. He had a few notes before him, to which he made occasional reference as he passed from one division to another. His lively inflections of voice, marked Scottish accent, and energetic gestures as he enforced the successive points which he wished the audience to comprehend were a novel and not unwelcome variation from the more usual formality of the presidential address. In the proceedings at the close reference was made to the fact that, though the President was not a Welshman, he had done his best to atone for that defect by marrying a Welshwoman. Ramsay in replying spoke of his love for Wales ; he knew almost every mountain-top in the Principality, he said, having either surveyed them with his own hands, or having superintended the surveys of them by others.

Towards the end of the year two marks of recognition of Ramsay's lifelong devotion to science were received by him. At the Anniversary of the Royal Society on the 1st December a Royal Medal was given to him 'for his long-continued and successful labours in geology and physical geography.' A few days later the University of Glasgow conferred on him the degree of LL.D.—thus towards the close of his career linking him by a new tie with his native town and the college with which he had so many pleasant early associations.

The second edition of the North Wales Memoir, which had involved so much labour both in the field and at the desk, was at last published at the close of 1881. In bulk it considerably exceeded the previous edition. A special interest attaches to it because it was its author's last Survey publication. As the year wore on it had become more and more evident that he must seek retirement from the endless cares of official life. In the course of the summer he made a round of farewell visits among his staff. I accompanied him through some parts of the centre of Scotland. He particularly wished to see some of the Highland lakes. So we made for Kenmore, and sailed up Loch Tay, and then by Lochs Vennachar and Achray to Loch Katrine and Loch Lomond. The scenery brought back early associations to him, and mingled with these reminiscences came the new interest which such scenery had for him in its bearing upon his doctrine of the glacial origin of lake-basins, and at the same time the sadness that arose from the feeling that he should probably never see these scenes again.

The British Association held its jubilee this year at York, where it had opened its career fifty years before. Ramsay, as the oldest surviving president of Section C, was asked to take the chair of that section on this occasion. He did so, and gave the address; but the effort was a great strain upon him, and he returned to Beaumaris to rest. It was definitely arranged that he should retire from his Government appointment at the end of the year.

To his old friend and colleague, Mr. Howell, he wrote : ' I feel grateful for the regret that our good fellows feel for my retirement. I regret it too very much, but in the words of the old ballad, " I'm weary wi' hunting and fain would lie down." I hope I may find contented rest in doing nothing but what I choose to do. The change, however, will be very great, even though the non-official intercourse should continue between us as fast as ever.'

He announced to his friends at Dolaucothi : ' On the 31st December I shall retire from the public service, and whether or not there will be another Director-General I do not know. Neither do I quite know how I shall enjoy doing nothing but what I choose to do ; but, on the whole, I think I shall manage very well. They have given me the highest possible pension, and that and our private incomes will enable us to live just as we have been accustomed to do ever since I became Director-General. . . . I think that, on the whole, I have been a " fortunate youth." One thing also pleases me, that I shall be able some time in 1882 to pay a visit to Dolaucothi the beloved, and to lie upon a bank where the wild thyme grows, and where oxlip and the nodding cowslip blows."

' My address to the geological section of the British Association at York last summer principally dealt with the progress of geology for the last fifty years. In my mind there is no doubt that it is, or was, the last address I shall ever give.

' The other day, as Louisa, Fanny, Dora, and I had arrived at the pudding stage of dinner, a franked letter arrived from Mr. Gladstone, informing me that at the instance of Lord Spencer, who is my official chief, I am required on Wednesday next to go down to Windsor by the 1.10 P.M. train in *levée* costume, and from the station, along with others, am to be transported to the castle to be knighted at three.'

On the afternoon of the 31st December Sir Andrew quitted his desk at the Jermyn Street Museum, and closed his long and honourable career as a civil servant.

For upwards of forty years he had given himself with his whole heart to the work of the Geological Survey, and he carried with him into his retirement the affectionate wishes of every member of the staff over which he had so long and so ably presided.

CHAPTER XI

RETIREMENT—SUMMARY OF CAREER

RELEASED from official life, and free to go where he
pleased in this country or to travel abroad, Sir Andrew
Ramsay looked forward to a few years of pleasant
rest and cheerful occupation in the pursuits that more
especially interested him. And to his friends the
regret at seeing him retire from active life was
tempered by the reflection that now at last he would
be able to work as he chose at those problems which,
by the pressure of his Survey duties and his engagements in London, he had been prevented from
thoroughly investigating. But time and over-exertion
had done their work upon him. His life henceforward
was marked by a calm, painless, and gradual decline.
His days for mental exertion were now at an end.

For a short time after his retirement he remained
in London, and had a fair enjoyment of life, usually
finding his way to the Athenæum Club in the course
of the afternoon, and having a game of billiards with
some friend there. When summer came, and brought
with it the time for retreating to Beaumaris, he was
able to take short walks and to indulge his favourite
pastime of steering a boat. He tried to do a little
writing on a geographical subject which he had
undertaken. As the autumn approached, the second

daughter was again ordered abroad for the winter, and
as Sir Andrew had such a love of mountains, it was
determined to move the whole family to St. Moritz.

In the early part of the season the place gave him
great pleasure, for he was able to take some fairly long
walks, and it was a delight to him to find himself once
more in the chain of the Alps.   But in the middle of
November the snow came, not to depart for the rest
of the winter.   Thereafter every walk involved a
slippery descent from the hotel and a slippery ascent
in returning.   The consequent fatigue became so great
that his pedestrian excursions were necessarily limited.
Sledge-driving proved equally impracticable, for the
keen frosty air induced severe pain round the glass
eye.   He was thus cooped up in the hotel, the most
practicable exercise available being obtained by pacing
up and down a covered verandah open to the air.

By the end of March the party was glad to take
flight by the Maloja to Chiavenna, and thence by
steamer on the Lake of Como to Cadenabbia.   Sir
Andrew enjoyed watching the pleasure which these
charming scenes gave to his daughters, who saw them
for the first time, and he seemed himself also to appreciate their beauty ; but the confinement of the preceding winter had left its mark upon him.   The old
spirit of happy comprehension of what he saw, and the
keen zest with which he scrutinised new physical
features and sought to interpret them, were now only
too visibly on the wane.   The route homeward included a halt of a fortnight at Venice, which he
thoroughly enjoyed, and another for the same length
of time at Pallanza, which also gave him much
pleasure.

The family returned to England in time to spend

the summer at Beaumaris, where Sir Andrew's diminished strength was painfully shown by his shortening walks. They went back, however, to the Continent in the autumn in order to pass the winter at Hyères. The journey homeward in May (1882) proved to be the last of Ramsay's experiences of foreign travel. He was well enough to be greatly interested in the Roman remains of Southern France. At Nîmes his antiquarian zeal was kindled by the grand amphitheatre and the baths and the Maison Carrée. He walked across the Pont de Gard by the old water channel. Farther north his ardour for the relics of the past was renewed in Auvergne, as he paced the mouldering rampart of the hill of Gergovia, and pictured to himself Cæsar's siege and the heroic defence of Vercingetorix. Fain would he have climbed the Puy de Dôme, but the weather prevented him from attempting it.

For the next two years Sir Andrew and his family came up to London for the winter, but he hardly ever went to the Athenæum, and was only able slowly to walk up and down the streets in the neighbourhood of his house in Cromwell Crescent. It was then resolved to break up the London home, in order that he might remain permanently at Beaumaris. There he continued for a time to enjoy the panorama of his own Caernarvonshire mountains, and was much more in the open air than he could have been in London. But from this time forward nothing could be done save to watch with sad and affectionate eyes the progress of the slow decline. His mind does not seem in these last years to have reverted often to his geological days; at least he seldom spoke of them. His memory would sometimes dwell on the long bygone days of his childhood.

He continued to read with delight the Waverley Novels, and the humour of Dean Ramsay's *Reminiscences of Scottish Life* never failed to call out his merry laugh. His general health remained good, but his strength, bodily and mental, seemed imperceptibly to ebb away.

Almost every fine day until 1891 he was wheeled out in a bath-chair and placed in some sunny, sheltered spot where he could watch the mountains and the sea, while his wife sat and read or worked beside him. These daily little journeys continued to give him great pleasure, until at last, in September of that year, his increasing weakness made them no longer possible. Eventually he was unable to bear the fatigue of rising and being dressed, so kept his bed until, on the 9th December, he passed gently away.

He was buried in the churchyard that surrounds the pretty little church of Llansadwrn, among his wife's people. The spot was a favourite one with him, for it commands on the one side a noble view of the whole range of the high grounds of North Wales from the Orme's Head, through the Snowdon group, down to the far Rivals, and on the other a wide sweep of the undulating plains of Anglesey. It was fitting that one who had loved Wales so ardently, who had spent the best years of his life there, and who had done more than any other writer to unravel at once its geology and its physical geography, should be laid to rest within view of the peak of Snowdon, and within sound of the rush of the tide through the Strait of Menai.

.   .   .   .   .

This memoir would be incomplete if it did not give some retrospect and summary of the work achieved in the lifetime which it has attempted to describe. It is too soon yet properly to appraise the ultimate value of this work in the general progress of science. But we may at least group Sir Andrew Ramsay's labours in the several categories under which they may be classified in order to form some conception of the general character and sum of his contributions to the geology of his time.

I. The department of Structural Geology comprises his earliest and his latest labours, beginning with his little pamphlet on Arran, and ending with the voluminous second edition of the monograph on North Wales. Between these two limits he accomplished a large amount of investigation directed towards the elucidation of the geological structure of Britain. In England his own share of this labour was for the most part merged in that of his colleagues. For, in his eagerness for the repute of the Survey as a body, he was careless of his individual fame. Undoubtedly his own greatest achievement is his mapping in North Wales, and more particularly the working out of the structure of the complicated and mountainous ground around Snowdon. It must be remembered by those who now examine the geology of that region that when Ramsay surveyed it the science of petrography hardly existed at all in England. He had no assistance from the microscope, and scarcely any from the chemical laboratory. He had to determine his rocks with no more help than could be given by a pocket-lens, and he was guided in this matter largely by the behaviour of the masses in the field. It should not, therefore, be matter for surprise that a geologist of to-day, coming with all the appliances of modern petrography, may be able to improve the nomenclature followed by Ramsay, and to show that he had been mistaken in some of his determinations, as where, for instance, he may have classed lavas as tuffs, or tuffs as lavas. The surprise ought rather to be that a man with only field-evidence and his geological instinct to guide him, should have succeeded in unravelling so admirably the complications of so difficult a region.

British geology lies under a deep obligation to Ramsay for the skill and insight with which he deciphered the relics of the older Palæozoic volcanoes. Without attempting to enter into the minutiæ of the mineralogical and chemical constitution of the rocks, he seized upon the salient features that illustrated ancient volcanic action, and he supplied, in the Survey Memoirs and in the Descriptive Catalogue of the Jermyn Street Museum, the first detailed and connected description of the different epochs of volcanic activity in the Silurian period in Britain.

On the maps and sections of the Geological Survey he expressed most of his results in structural geology, and it may be fearlessly asserted that at the time of their appearance these publications were unsurpassed for clearness, beauty, and accuracy. Even where the mapping was mainly the work of his colleagues, it usually had the benefit of help from his skilful hand and sound judgment. His Geological Map of England and Wales, reduced from the Survey sheets and other sources of information, is still the most useful small map of the kingdom, and his general Geological Map of the British Isles is a convenient compendium of British geology.

II. In Stratigraphy much of Ramsay's work is so intimately bound up with his labours in structural geology as to be hardly separable. But his two presidential addresses to the Geological Society mark a distinct epoch in stratigraphical work. Darwin had dwelt upon the imperfection of the Geological Record. Ramsay proceeded to indicate the historical meaning of this imperfection. He pointed out the various breaks in the succession of the stratified formations of Britain, and by his wide practical knowledge of the subject, gave it a clearness and significance which it had not before been suspected to possess. He showed that these breaks sometimes consist of actual unconformabilities, arising from disturbance and denudation, and demonstrating a long lapse of time unrecorded by stratified deposits ; while in other instances they are marked by no visible discontinuity of the stratification, but by a sudden and more or less marked change in the fossils characteristic of two apparently consecutive formations. His careful tabulated lists of genera and species that pass from one formation to another finally annihilated the long-lived delusion that each geological system was complete in itself, and was separated, by a general destruction and re-creation of life, from the formation that succeeded it. Accepting Darwin's views on the origin of species, he argued that the relative lapse of time between different formations might be determined by the greater or less distinction between them as regards their organic contents. By this line of argument he was led to the novel and suggestive conclusion that periods of time, of which there was in the geological record of Britain no stratigraphical chronicle, might be much longer than those which were represented by stratified formations.

These doctrines were by far the most important which had been taught in regard to the principles of stratigraphy since these principles were first determined by the discoveries of William Smith.

III. Connecting his Stratigraphical with his Physiographical researches comes the series of papers in which he discussed the former existence of Continents, or of terrestrial conditions, during the deposition of the geological record. He dwelt especially upon the red colour of certain formations, their barrenness in organic remains, the proofs of the occurrence of traces of land animals and plants in them, and the similarity presented by them to the deposits of salt lakes or inland seas. In this way he tried to restore in some degree the physical geography of ancient periods of the earth's history. He attempted also, from the same kind of reasoning, to estimate the relative value of the old continental periods, and came to the conclusion that the period which began with the Old Red Sandstone and closed with the New Red Marl may have been comparable to all the time that has elapsed from the beginning of the deposition of the Lias down to the present day.

IV. In Physiography Ramsay's work was abundant, as well as remarkably original and important. It may be grouped in three subdivisions : (1) Denudation in General ; (2) The History of River-valleys ; and (3) The Results of the Operations of Ice.

(1) The early paper on the Denudation of South Wales, published in 1846, in the first volume of the *Memoirs of the Geological Survey,* was undoubtedly the most important essay on the subject which up to that time had appeared. Much had previously been written on the question of denudation, but it was of the

vaguest nature. It was Ramsay's merit that he based his discussion upon the results of careful surveying. He had traced out the structure of a complicated geological region, and was able to show what should have been the form of the surface had it depended entirely on geological structure. He was thus in a position to demonstrate how much material had been removed by denudation, and how far the process of removal had been guided by geological structure. It is true, as he himself afterwards confessed, that at that time he assigned too much power to the sea, and too little to the subaerial agents, in the lowering of a mass of land. But his exposition of the old base-level of ancient erosion, or 'plain of marine denudation,' as he called it, will ever be a classical study in geological literature.

Subsequently, as he realised more and more how mighty had been the action of rain, frost, rivers, glaciers, and other subaerial forces in carving the surface of the land, he came boldly forward to take the lead among British geologists in enforcing this doctrine.

(2) During the last ten years of his official life the physical history of river-valleys exercised a peculiar fascination on Sir Andrew's mind. The subject had for many years engaged his attention, but not until the appearance in 1862 of his friend Jukes's remarkable memoir on the river-valleys of the south of Ireland did he realise how the problem might be satisfactorily attacked. He was led to the conclusion that the denudation of the Weald had been effected by subaerial waste, and that the cause of the flow of the rivers, from that central low tract through the encircling rim of chalk downs, was to be sought in the ancient topo-

graphy of the region, when the streams descended from a central, still unremoved dome or ridge of chalk. Extending this process of reasoning, he afterwards discussed the main causes whereby the rivers of England had been led to flow in the courses which they now follow. There was undoubtedly a good deal of speculation in this discussion, but his treatment of the subject was full of suggestiveness, and pointed out the direction in which, with perhaps a larger array of facts, the question might eventually be solved.

Subsequently he attacked the history of individual rivers, working it out in more detail along the same lines as he had already followed. In this manner he traced the successive stages which, in his opinion, had led to the excavation of the present valley of the Rhine, showing that in Miocene time the flow of the drainage between the Black Forest and the Vosges had been from north to south, or towards the great hollow lying to the north of the Alps, that subsequent disturbance and elevation of the Alpine chain tilted the ground in such a manner that the drainage was reversed, and the streams from the tract of the Alps were collected into a river which found its way northward, and gradually excavated the valley and gorge in which the present Rhine still flows. Though it cannot be demonstrated that such have been the successive stages in the history of the course of this river, the available evidence makes Ramsay's explanation highly probable.

The later application of the same principles of interpretation to the history of the valley of the Dee in Wales led him into still ampler fields of speculation, because dealing with a vaster and dimmer geological past. He could not claim to have proved every step

in his chain of argument, but he undoubtedly propounded a method whereby, if such questions are
capable of solution, they may most advantageously
be attempted.

(3) It was in his researches among the traces of
ancient glaciers and ice-sheets that Sir Andrew accomplished his most original physiographical work. His
demonstration of the occurrence of evidence of two
glaciations in Wales was an important step in the
elucidation of the history of the Ice Age, for he showed
that after a general glaciation of the Welsh hills and
valleys, and the deposition of the drift upon them, a
later time came when the ice existed only as local
glaciers in the valleys among the higher mountains.

But his name will be most widely known for his
theory of the Glacial Origin of certain Lake-basins.
This theory has been warmly attacked and as vigorously defended. The contest regarding it still continues, though more than thirty years have passed
since it was published. This is not the place for a
review of the voluminous arguments that have been
adduced for and against the theory. If we look upon
the doctrine as promulgated by its author, and not in
the extravagant form in which it sometimes appears in
the hands of too zealous partisans, we must admit that
Ramsay was the first to call attention to the remarkable fact that lakes are especially numerous in the
glaciated tracts of the northern hemisphere. Taking
the rock-basin lakes on which he based his doctrine of
glacial erosion, it is a fact that while they are prodigiously abundant in glaciated tracts like the gneisses
of Canada, Scandinavia, Finland, and Scotland, they
either do not occur, or are excessively rare, outside of
ice-worn areas. It is likewise true that these rock-

basins have once been filled with ice, for *roches moutonnées* occur round their margins and rise from their bottoms. There can be no doubt also that the ice which filled them was in motion, for the rocks that enclose them are scored and polished, and the direction of the striæ shows that the ice descended into the basins at their upper end and ascended from them at the lower. Ramsay went farther, and insisted that the hollows themselves had actually been dug out by the moving ice. I have myself no doubt that he was essentially right in this contention. That there may be difficulty in the universal application of his doctrine will be readily admitted, and was fully recognised by himself. He carefully guarded himself by the very title of his original paper, 'On the Glacial Origin of *Certain* Lakes,' from being supposed to have one explanation for all sheets of fresh water over the surface of the globe. But that the lakes in glaciated regions are connected in origin with the general denudation of the regions in which they lie is a fact which few geologists who have carefully mapped the rocks around these water-basins will dispute. And the only agent known to us to be capable of the kind of erosion which would produce such basins is land-ice. On any other hypothesis yet proposed the lake-basins are not only unintelligible, but contradictory to all that is now well ascertained regarding the progress of denudation and the influence of geological structure upon topography.

In connection with Sir Andrew Ramsay's glacial work, reference should be made here to his papers on the evidence for the existence of ice in Palæozoic time. The Permian examples cited by him were certainly striking, but the general feeling among geologists seems to be that the evidence is not convincing. The cases from the Lower Silurian, Old Red Sandstone, and Carboniferous formations are still less conclusive.

As a contribution to Physiography his volume on *The Physical Geology and Geography of Great Britain* is worthy of special mention. It puts in clear and untechnical language the evidence on which geology proceeds to trace the bygone history of a terrestrial region. Unfortunately, the last edition was brought out by him when his powers had already begun to fail, and he was led to weight the book by the addition of various already published papers and essays which, though excellent for the purposes for which they were written, were somewhat out of place in his otherwise charming little volume.[1]

V. Sir Andrew made few contributions to the literature of the History of Geology. His two inaugural lectures at University College presented a rapid sketch of the leading features in the progress of geological research up to this century, and his last address as President of Section C of the British Association at York in 1881 gave an outline of the advance of geology during the fifty years preceding that date.

He was a thorough uniformitarian in geology. Having early imbibed his theoretical views from Lyell's *Principles*, he maintained them to the end, and took occasion when he was President of the British Association at Swansea, and consciously approaching the end of his active career, to proclaim them as a last declaration of faith to his contemporaries.

VI. In this retrospect of the literary and other work which Sir Andrew Ramsay accomplished, reference should not be omitted to his contributions to the

---

[1] A new edition of this book is now in preparation under the able editorship of Mr. H. B. Woodward of the Geological Survey.

Saturday Review. Sometimes these were criticisms
of publications, and in that case often took as their
subject the maps and memoirs of some Colonial
Survey. He was thus enabled to do signal service
to his friend Logan, then struggling with the Canadian
Philistines, who saw no practical good in geological
work of any kind. But now and then he let his
fancy loose in the pages of the *Saturday*, and treated
geological and other topics with the light playfulness
so characteristic of his talk, but in which the nature
of his official writings hardly allowed him to indulge
in print.

VII. But it is not by the visible amount of published
work that we can rightly estimate the extent of Sir
Andrew Ramsay's influence in promoting the advance
of his favourite science. For nearly thirty years he
was a teacher of geology. Year by year a fresh band
of young men came to listen to him, and to carry the
fruits of his instruction to all parts of the world.
Season after season he lectured to working men, who
flocked in hundreds to hear him. His lectures were
not written out, but delivered from notes, and were
always kept up to the latest conditions of the science.
Many a time some new deduction that had been
simmering in his mind for a while would be communicated first of all to his students. In the debates at
the Geological Society, also, he would often make
known some fresh observation, or some novel presentation of known facts, or some suggestive speculation
which had recently taken shape in his mind. Indeed,
much of his work was published only in this way,
for writing became increasingly irksome to him ; but
in the excitement of lecturing or of discussion he
would pour out from his full stores of information,

and taking his audience into his confidence, would flash out new views that he had never communicated to any one before. There was always something remarkably suggestive in his lectures. He loved to put broad and striking views of geological principles and theory before his audience, and sought thus to excite an interest that would search for detail, rather than to weary it by dwelling on the detail himself. He spoke with a good deal of facial expression, his brow sometimes wrinkling with his earnestness to make a point clear, and again beaming with a kindly smile as he was enjoying his discourse, and felt that he was carrying his auditors with him. The working men used to crowd round his table at the end of a lecture to ask questions, and one of them once said to him, 'You are the best lecturer I ever heard in my life ; and you always look so happy in it.'

There was another form of instruction less palpable perhaps than that communicated in formal prelections, but not less valuable—the practical training which he gave to his men on the staff of the Geological Survey. Those who have enjoyed that training look back upon it as one of the privileges of their lives. The influence of his example was contagious. A district which had seemed hopelessly entangled and insufferably dull, after a visit from the Director came to be seen in a new light. Its very difficulties grew to be centres of attraction, and its dulness was changed into freshened interest. That this influence has not been without effect in the higher education of the country will be seen from the number of men trained under Ramsay who have held or now hold University chairs or other educational appointments in this country and in the colonies.

But above and beyond the impress of his scientific achievements Sir Andrew Ramsay's high position among his contemporaries was largely determined by his individual personality. His frank manly bearing, his well-cut features beaming with intelligence and with a sweet childlike candour, his ready powers of conversation, his wide range of knowledge, his boyish exuberance of spirits, his simplicity and modesty of nature, his sterling integrity, perfect straightforwardness, and high sense of duty, his generous sympathy and untiring helpfulness, marked him out as a man of singular charm, and endeared him to a wide circle of friends who, while they admired him for his genius, loved him for the beauty and brightness of his character.

APPENDIX

I.—List of Papers, Books, etc., by Sir Andrew C. Ramsay, F.R.S.[1]

1840

'Notes taken during the Surveys for the Construction of the Geological Model Maps and Sections of the Island of Arran. *Rep. Brit. Assoc.* for 1840 [1841]. Sections, p. 92.

1841 and previous years

'The Geology of the Island of Arran, from original survey.' 8vo. Glasgow [1841].

1846

'On the Denudation of South Wales and the adjacent Counties of England.' *Mem. Geol. Survey*, vol. i. pp. 297-335.

1847

'On the Causes and Amount of Geological Denudations.' A discourse at the Royal Institution. *Athenæum*, 27th March 1847, p. 342.
'On the Origin of the existing Physical Outline of a portion of Cardiganshire.' *Rep. Brit. Assoc.* for 1847 [1848]. Sections, pp. 66, 67.

1848

'Sketch of the Structure of Parts of North and South Wales.' (With W. T. Aveline.) *Quart. Journ. Geol. Soc.* vol. iv. pp. 294-297.

---

[1] The years put in separate lines are generally those in which the work was done; the years placed within square brackets give the dates of publication.

'Passages in the History of Geology.' Two Lectures at University College, London. 8vo. London, 1848, and second lecture 1849.

'On some points connected with the Physical Geology of the Silurian district between Builth and Pen-y-bont, Radnorshire.' *Rep. Brit. Assoc.* for 1848 [1849]. Sections, p. 73.

1850

'On the Geological Phenomena that have produced or modified the Scenery of North Wales.' Friday evening discourse at Royal Institution. *Athenæum*, 6th April 1850, p. 377.

'On the Geological Position of the Black Slates of Menai Straits, etc.' *Rep. Brit. Assoc.* for 1850 [1851]. Sections, p. 102.

1852

'On the Science of Geology and its Applications.' Introductory Lecture to the Course of Geology, School of Mines, Session 1851-52. *Records School of Mines*, vol. i. pp. 81-102.

'On the Superficial Accumulations and Surface-markings of North Wales.' *Quart. Journ. Geol. Soc.* vol. viii. pp. 371-376.

1853

'On the Physical Structure and Succession of some of the Lower Palæozoic Rocks of North Wales and part of Shropshire,' with Notes on the Fossils by J. W. Salter. *Quart. Journ. Geol. Soc.* vol. ix. pp. 161-179.

1854

'On the Geology of the Gold-bearing District of Merionethshire, North Wales.' *Quart. Journ. Geol. Soc.* vol. x. pp. 242-247.

'On the former probable existence of Palæozoic Glaciers.' *Rep. Brit. Assoc.* for 1854 [1855]. Sections, pp. 93, 94.

'On the thickness of the Ice of the ancient Glaciers of North Wales, and other points bearing on the Glaciation of the Country.' *Rep. Brit. Assoc.* for 1854 [1855]. Sections, pp. 94, 95.

'On the occurrence of Angular, Subangular, Polished, and Striated Fragments and Boulders in the Permian Breccia of Shropshire, Worcestershire, etc. ; and on the Probable Existence of Glaciers and Icebergs in the Permian Epoch.' *Quart. Journ. Geol. Soc.* vol. xi. pp. 185-205.

1857

'On certain peculiarities of Climate during part of the Permian Epoch.' *Proc. Roy. Inst.* vol. ii. pp. 417-421.

1858

'The Physical Structure of Merionethshire and Caernarvonshire.' *Geologist*, vol. i. pp. 169-174.

'On the Geological Causes that have influenced the Scenery of Canada and the North-eastern Provinces of the United States.' *Proc. Roy. Inst.* vol. ii. pp. 522-524.

'Geology of Parts of Wiltshire and Gloucestershire.' (With W. T. Aveline and E. Hull.) *Geol. Survey Mem.* 8vo, London, 1858.

'Geological Surveys in Great Britain and her Dependencies.' *Saturday Review*, 3rd July, pp. 8-10.

'Descriptive Catalogue of the Rock Specimens in the Museum of Practical Geology.' (With H. W. Bristow, H. Bauerman, and A. Geikie.) 8vo. London, 1858; 2nd edit., 1859; 3rd edit., 1862.

1859

'On Some of the Glacial Phenomena of Canada and the North-eastern Provinces of the United States during the Drift-Period.' *Quart. Journ. Geol. Soc.* vol. xv. pp. 200-215.

'Geology of Pennsylvania.' *Saturday Review*, 30th April and 28th May, pp. 530, 531, 558-560.

'Beach Rambles.' (Review of J. G. Francis's 'Beach Rambles in Search of Sea-side Pebbles and Crystals.') *Saturday Review*, 12th November, pp. 585, 586.

'Geological Map of England and Wales.' (Stanford.) 1859. 3rd edit., 1872; 4th edit., 1877; 5th edit., 1881.

1860

'The Old Glaciers of Switzerland and North Wales.' 8vo. London. (Printed in *Peaks, Passes, and Glaciers*, by the Alpine Club. 8vo. 1859.)

'Geological Surveys of Canada and New Zealand.' *Saturday Review*, 11th August, pp. 174-176.

1861

'Lyell and Tennyson.' *Saturday Review*, 22nd June, pp. 631, 632.
'Kensington Gardens.' *Saturday Review*, 29th June, pp. 668, 669.

1862

'On the Glacial Origin of certain Lakes in Switzerland, the Black Forest, Great Britain, Sweden, North America, and elsewhere.' *Quart. Journ. Geol. Soc.* vol. xviii. pp. 185-204.
'The Excavation of the Valleys of the Alps.' *Phil. Mag.* vol. xxiv. pp. 377-380.

1863

'Breaks in Succession of the British Palæozoic Strata.' Address to Geological Society. *Quart. Journ. Geol. Soc.* vol. xix. pp. xxix–lii.

1863-78

'The Physical Geology and Geography of Great Britain.' 8vo. London. 2nd edit., 1864; 3rd edit., 1872; 4th edit., 1874; 5th edit., 1878.

1864

'Notes accompanying Translation of Paper on "The Sahara, and its Different Types of Deserts and Oases,"' by E. Desor. *Geol. Mag.* vol. i. pp. 27-34.
'The Breaks in Succession of the British Mesozoic Strata.' Address to Geological Society. *Quart. Journ. Geol. Soc.* vol. xx. pp. xl–lx.
'On the Erosion of Valleys and Lakes: a Reply to Sir Roderick Murchison's Anniversary Address to the Geographical Society.' *Phil. Mag.* vol. xxviii. pp. 293-311.
'The Geology of Canada.' *Saturday Review*, 26th March, pp. 383, 384.
'Changes of Climate.' *Saturday Review*, 14th May and 11th June, pp. 591, 592, 719-721.

1865

'The Ice-drifted Conglomerates of the Old Red Sandstone.' *Reader*, 12th August, p. 186.

'On the Eozoon and the Laurentian Rocks of Canada.' *Proc. Roy. Inst.* vol. iv. pp. 374-377.

'Sir Charles Lyell and the Glacial Theory of Lake Basins.' *Phil. Mag.* vol. xxix. pp. 285-298.

1866

'The Geology of North Wales.' With Appendix on Fossils by J. W. Salter. *Geol. Survey Mem.* 8vo. London, 1866. 2nd edit. 1881.

'Address to the Geological Section of the British Association at Nottingham.' *Rep. Brit. Assoc.* for 1866 [1867]. Sections, pp. 46-50.

1868

'Review of 1st vol. of Lyell's " Principles of Geology." ' *Saturday Review*, 11th and 18th April.

1869

'Evidence given before the Royal Commission on Water-Supply.' *Report*, pp. 104-107. Folio. London.

'Report on Ice as an Agent of Geologic Change. (With O. Torell and H. Bauerman.) *Rep. Brit. Assoc.* for 1869 [1870], pp. 171-174.

1871

'On the Recurrence of Glacial Phenomena during great Continental Epochs.' *Nature*, vol. v. pp. 64, 65.

'On the Physical Relations of the New Red Marl, Rhætic Beds, and Lower Lias. *Quart. Journ. Geol. Soc.* vol. xxvii. pp. 189-199.

'On the Red Rocks of England of older date than the Trias.' *Quart. Journ. Geol. Soc.* vol. xxvii. pp. 241-256.

'Report on the Probability of finding Coal under the Permian, New Red Sandstone, and other superincumbent Strata.' *Roy. Comm. on Coal*, vol. i. pp. 119-145. Also *Evidence*, vol. ii. pp. 422-424, 449, 463-488. Folio. London.

1872

'On the River-courses of England and Wales.' *Quart. Journ. Geol. Soc.* vol. xxviii. pp. 148-160.

'The River Po.' *Macmillan's Magazine*, vol. xxvii. pp. 125-129.
('Les Inondations en Italie.') Il Fiume Po. *Bolletino del Club Alpino Italiano*, vol. vi. 8vo. Turin, 1873. (Translated from *Macmillan's Magazine*, December 1872.)

1872-1881

'Annual Reports of the Director-General of the Geological Survey of the United Kingdom, the Museum of Practical Geology, the Royal School of Mines, and the Mining Record Office, for the years 1871 to 1881.' *Reports of Science and Art Department.*

1873

'On Old Continents.' *Proc. Roy. Inst.* vol. vii. pp. 32-34. (Translated into French in the *Revue Scientifique.*)
'Report on the Exploration of Brixham Cave,' conducted by a Committee of the Geological Society. (Report prepared by J. Prestwich, agreed and approved by G. Busk, R. Godwin-Austen, and A. C. Ramsay.) *Phil. Trans.* vol. clxiii. pp. 471-572.
'The History of Great Britain.' *Saturday Review*, 25th January, pp. 110, 111.

1874

'On the Comparative Value of certain Geological Ages (or groups of Formations) considered as Items of Geological Time.' *Proc. Roy. Soc.* vol. xxii. pp. 334-343.
'Article "Geology" in Blackie's "The Popular Encyclopædia."' 8vo. London.
'On the Physical History of the Rhine.' *Proc. Roy. Inst.* vol. vii. pp. 279-288.
'The Physical History of the Valley of the Rhine.' *Quart. Journ. Geol. Soc.* vol. xxx. pp. 81-95.

1875

'The Pre-Miocene Alps, and their subsequent Waste and Degradation.' *Proc. Roy. Inst.* vol. vii. pp. 455-457.
'Geological History of some of the Mountain Chains and Groups of Europe.' Lectures at the Royal School of Mines. *Mining Journal*, vol. xlv. pp. 57, 79, 106, 135, 162, 191.

'General Instructions for Observations in Geology.' (With J. Evans.) In 'Manual of the Natural History, etc., of Greenland and the neighbouring Regions'; prepared for the use of the Arctic Expedition of 1875. 8vo. London. Pp. 68-77.

'Evidence given before Civil Service Inquiry Commission: and Memorandum on the Geological Survey.' *Second Report of Commission, with Appendix.* Folio. London. Pp. 43-46, 70-72.

'Orographical Series of Wall Maps.' (Stanford's.) British Isles, Europe, etc.

1876

'How Anglesey became an Island.' *Quart. Journ. Geol. Soc.* vol. xxxii. pp. 116-122.

Note appended to paper by William Ramsay, 'On the Influence of Various Substances in Accelerating the Precipitation of Clay suspended in Water.' *Quart. Journ. Geol. Soc.* vol. xxxii. p. 132.

'On the Physical History of the Dee, Wales.' *Quart. Journ. Geol. Soc.* vol. xxxii. pp. 219-229.

'The Origin of Lake-Basins.' (A Letter.) *Geol. Mag.* Decade II. vol. iii. pp. 136-138.

1877

'The Origin and Progress of the Geological Survey of the British Isles, and the Method on which it is conducted.' In *Science Conferences.* Conferences held in connection with the Special Loan Collection of Scientific Apparatus, vol. ii. pp. 364-380. 8vo. London.

'Report on the Question of the Supply of Fresh Water to the Town and Garrison of Gibraltar.' Folio. London.

'The Existence of Coal beneath the New Red and Permian Strata.' *Proc. Dudley Geol. Soc.* vol. iii. pp. 35-37.

1878

'The Geology of Gibraltar and the Opposite Coast of Africa; and the History of the Mediterranean Sea.' *Proc. Roy. Inst.* vol. viii. pp. 594-601.

'Geological Map of the British Isles.' (Stanford.) 1878.

'On the Geology of Gibraltar.' (With J. Geikie.) *Quart. Journ. Geol. Soc.* vol. xxxiv. pp. 505-541.

1879

'Discussion at Annual Conference on National Water-Supply, etc.'
Journ. Soc. Arts, vol. xxvii. p. 159.

1880

'On the Recurrence of certain Phenomena in Geological Time.'
(Presidential Address to the British Association.) *Rep. Brit.
Assoc.* for 1880, pp. 1-22.

1881

'On the Origin, Progress, and the Present State of British Geology,
especially since the first meeting of the British Association at
York in 1831.' Address to Section C. *Rep. Brit. Assoc.* for
1881 [1882], pp. 605-608.

1883

'Notes on the Geology of St. David's.' (1842-43.) In paper 'On
the Supposed Pre-Cambrian Rocks of St. David's,' by A.
Geikie. *Quart. Journ. Geol. Soc.* vol. xxxix. pp. 263, 264.

1885

'Stanford's Compendium of Geography—Europe.' (With F. W.
Rudler, G. G. Chisholm, and A. H. Keane.) 8vo. London.

II.—LIST OF GEOLOGICAL SURVEY MAPS (ONE-INCH SCALE), PARTS
OF WHICH WERE SURVEYED BY SIR A. C. RAMSAY.[1]

England and Wales

Sheet 19.   (1845) Mendip Hills, Bath.
  ,,   35.   (1845) Bristol, Chepstow, Cotteswold Hills.
  ,,   38.   (1845) Pembroke, Milford Haven.
  ,,   40.   (1845) St. David's.
  ,,   41.   (1845) Caermarthen, Llandeilo, Llandovery.
  ,,   43. S.E. (1845) Woolhope, Malvern, Ledbury.
  ,,   53. N.W. (1855) Coventry, Rugby, Leamington.
  ,,   54. N.W. (1852) Kidderminster, Bromsgrove.

---

[1] The original dates of publication are given, but new editions of some of the
Sheets have been published.

Sheet 56. N.W. (1850) Rhayader.
" 56. N.E. (1850) Clun, Knighton, Presteign.
" 56. S.W. (1850) Builth.
" 56. S.E. (1850) New Radnor.
" 57. N.W. (1848) Cardiganshire coast.
" 57. N.E. (1848) Aberystwith.
" 57. S.W. (1848) Aberavon, Aberforth.
" 57. S.E. (1848) Lampeter, Tregaron.
" 58. (1850) Cardiganshire coast.
" 59. N.E. (1850) Barmouth, Dolgelly, Cader Idris.
" 59. S.E. (1848) Machynlleth, Aberdovey, Plinlimmon.
" 60. N.W. (1851) Dinas Mowddwy, Llanfair.
" 60. N.E. (1850) Welshpool.
" 60. S.W. (1850) Llanidloes.
" 60. S.E. (1850) Montgomery, Bishops Castle.
" 61. N.W. (1855) Shrewsbury.
" 61. N.E. (1855) Wellington, Wrekin.
" 61. S.W. (1850) Church Stretton, Longmynd.
" 62. N.W. (1852) Cannock Chase.
" 62. S.W. (1852) Wolverhampton, Dudley.
" 62. N.E. (1856) Lichfield, Tamworth.
" 62. S.E. (1855) Birmingham.
" 71. N.W. (1855) Belper, Wirksworth.
" 72. N.E. (1852) Ashbourn.
" 74. N.W. (1850) Corwen, Pentre Voelas.
" 74. N.E. (1850) Crewe, Nantwich.
" 74. S.W. (1850) Bala, Hirnant.
" 74. S.E. (1850) Oswestry.
" 75. N.W. (1850) Nevin, Caernarvonshire.
" 75. N.E. (1851) Snowdon, Tremadoc.
" 75. S.E. (1851) Harlech, Portmadoc.
" 77. N. (1852) Holyhead Island.
" 78. N.W. (1852) Anglesey, part of Holyhead.
" 78. N.E. (1852) Beaumaris, Conway.
" 78. S.W. (1852) Anglesey, Caernarvon.
" 78. S.E. (1852) Bangor, Llanberis.
" 79. N.W. (1850) Little Orme's Head.

Scotland

" 33. (1860) Dunbar, Haddington.

III.—List of Horizontal Sections of the Geological Survey
drawn by Sir Andrew C. Ramsay.

Sheet 1

No. 1. Section from Porth-wyn, St. Bride's Bay, near Solva, Pem-
brokeshire, to the north cliff of Ynys-y-Barry ; on a line 100°
W. of N. from Porth-wyn to the cross-roads near Shyvog
Common, and 5° E. of N. from thence to the sea. By A. C.
Ramsay and W. T. Aveline.

No. 2. Section across Pembrokeshire, from South to North. From
St. Gowan's Head to Lanstadwell, from Lanstadwell to Tref-
garn Rock, and thence to Dinas Head, near Fishguard. By A.
C. Ramsay and W. T. Aveline. (1845. Revised 1857.)

Sheet 3

No. 1. Section from Cerrig-dwfn to Mynydd-bancy-ffair, near
Llandeilo. By A. C. Ramsay.

No. 3. Section from the Black Mountain, near Llangadoc, to
Cefnllwyn hir, Caermarthen. By H. T. De la Beche, A. C.
Ramsay, and W. T. Aveline. (1844.)

Sheet 4

No. 4. Section from Mynydd Bwlch-y-groes, Brecknock to Craig-ddu,
Cardigan Bay. By A. C. Ramsay and W. T. Aveline. (1845.
Revised 1858.)

Sheet 5

No. 1. Section across the Old Red Sandstone and Silurian Rocks,
from the Black Mountain Range S.E. of Glasbury to Allt-wen,
Cardigan Bay, near Aberystwith. By A. C. Ramsay, T. E.
James, and W. T. Aveline. (1845. Revised 1858.)

Sheet 6

No. 2. Section across the Silurian Rocks, from Gwaun Ceste to
Rhiw Gwraidd, Radnor. (1845. Revised 1858.)

Sheet 13

No. 1. Section from Edge Hill, Forest of Dean, to Taynton House. (N.D.; about 1845.)

Sheet 14

No. 2. Section from the Great Western Railroad, near Saltford Station, to the Box Valley near Slaughterford. (1845.)

Sheet 15

No. 1. Section from Ridge Barn Hill, near Castle Cary, Somerset, to Jay Hill, near Bitton, Gloucester. By H. T. De la Beche, D. H. Williams, and A. C. Ramsay. (1845.)

Sheet 27

Section across Radnor Forest and Hanter Hill, to the Old Red Sandstone near Fern Hall, South of Kington, Herefordshire. By A. C. Ramsay, H. W. Bristow, and W. T. Aveline. (1852.)

Sheet 28

Section from Llanfair-is-gaer, Menai Straits, over the Cambrian and Silurian Rocks of Dinas, Snowdon, Cynicht, Moel-wyn, Corsgoch (near Trawsfynydd), Aran Mowddwy, and Newtown, Montgomeryshire, and the Upper Silurian Rocks and Old Red Sandstone of Clun Forest, Bucknall, Wigmore Valley, Orleton, etc., near Ludlow. (1853.)

Sheet 29

Section from Craig das Eithin across Y Dduallt, Aran Mowddwy, and Cwledog, Merionethshire. By A. C. Ramsay, A. R. C. Selwyn, and W. T. Aveline. (1853.)

Sheet 31

Section from the Suspension Bridge, Menai Straits, across Y-Glyderfawr, Moel Siabod, Bala Lake, etc. (1854.)

Sheet 40

No. 1. Section from Porth Llanlliana, across Anglesey to the Menai Straits at Llanidan.

No. 2. Section from Carmels Point to the north shore of Church Bay, N.W. of Anglesey.

No. 3. Section across Anglesey from Point Ælianus to the Menai Bridge by Traeth Dulas and Pentraeth. (With explanation, 1857.)

Sheet 44

No. 2. Section through Tan-y-Castell, Cefn-y-Fedw, etc. (Account of Silurian, Old Red Sandstone, and Carboniferous Rocks in 'Explanation,' 1859.)

Sheet 45

No 1. Section across the Upper Silurian Rocks and Coal-measures of Coalbrook Dale, etc. (Notes in 'Explanation,' 1859.)

Sheet 50

Section from near Cleobury Mortimer to Nuneaton, Forest of Wyre, Clent Hills, etc. (Notes in 'Explanation,' 1859.)

Sheet 54

No. 1. Section from N.W. to S.E., across the Wrekin, Coalbrook Dale Coalfield, Shropshire, and the New Red Sandstone of Beckbury, Pattingham, and Oreton Hill, to Baggeridge Wood, South of Wolverhampton, Staffordshire. By A. C. Ramsay, D. H. Williams, and E. Hull (1858). (With explanation, 1859.)

Sheet 58

No. 2. Section from W. to E. from Cluddley, near Wellington, across the Coalbrook Dale Coalfield, the Permian Rocks, and New Red Sandstone, near Shiffnal and Brewood to the Western Border of the S. Staffordshire Coalfield S. of Cannock. By A. C. Ramsay, D. H. Williams, and E. Hull. (1860.)

IV.—LIST OF VERTICAL SECTIONS OF THE GEOLOGICAL SURVEY
DRAWN BY SIR ANDREW C. RAMSAY.

Sheet 12

Sections illustrative of the passage of the Old Red Sandstone into
the Carboniferous Limestone in South Wales and South-Western
England.

No. 2. Skrinkle Haven, Pembrokeshire.

Sheet 15

No. 1. Section through the Silurian strata of May Hill, Gloucestershire.

INDEX

Geological Survey of Great Britain, founded by H. T. De la Beche, 34; first beginnings of in 1832, 36, 37; under Board of Ordnance, 37; publication of early maps, 37, 38, 39; connection of with Geological Society, 37, 120; character of first maps, 39, 205, 206; field-work extended to South Wales, 42, 43; duties of officers of, 43; military uniform of, 45; life of, in the field, 45; discomforts and risks in work of, 46; palæontologist appointed, 57; influence of climate on work of, 59; winter work of, 60; transferred to the Office of Works, 61, 65; augmentation of staff, 66; publishes *Memoirs*, 67, 84; account of by Leonard Horner, 68; system of accounting in, 69, 168; duties of Local Director of, 71; work of in Ireland, 72; work of in South Wales finished, 75; work of in North Wales begun, 79; in progress in the Dolgelli region, 81; appealed to by Murchison, 93; revision work of (*see* 'Revision'); preliminary traverses in Snowdon region, 105; section-running by, 113; progress of up to 1848, 120; first sketch of work done in North Wales, 125; first publication of results of work in North Wales, 125; regulation as to publication of results obtained by officers of the, 126; work in Snowdon region, 133, 148, 153, 161; legal powers of trespass, 133; annual dinners of, 142, 160, 175, 241; songs by members of, 142, 160, 161, 176, 241, 242; head office removed to Jermyn Street, 174; transferred to Department of Science and Art, 207; extended to Scotland, 209, 219; progress of in England up to 1854, 211; condition of staff in 1854, 212; six-inch Ordnance Survey maps used by, 215, 217; death of De la Beche, 227; Murchison appointed as his successor, 228; explanatory memoirs for the maps organised, 259; reorganisation and enlargement of staff in 1867, 290-293; agronomical work, 293; progress of the service up to 1872, 314; duties of Director-General, 315; preparation of Annual Report,

321; revision of maps of Wales (*see* 'Revision'); conference of Director-General with Directors for England and Scotland, 342; Sir A. C. Ramsay resigns his position on the staff, 348
Geology, Structural, Ramsay's contributions to, 355
—— Stratigraphical, Ramsay's work in, 357
—— Physiographical, Ramsay's researches in, 358
—— history of, 349, 363
—— Museum of Economic (*see* Museum)
Gergovia, 353
Giant's Causeway, 342
Gibbs, R., 66, 83, 105, 108, 135, 136, 137, 147, 212
Gibraltar, 337, 338, 344
Glacial geology, Ramsay's first reference to, 64; his first lessons in, 137; early incredulity regarding, 138; Ramsay's first public declaration relating to, 160; progress of Ramsay's views regarding, 167, 168, 170, 177, 193, 198, 202, 219, 251, 253, 254, 268, 326, 337, 345, 346; summary of his contributions to, 361
—— periods, successive, 198, 219, 228, 322, 326, 362
Gladstone, Mr. W. E., 349
Glas lyn, 135
Glasgow, Chemical Society of, founded, 2
—— Philosophical Society of, foundation of, 2
—— University confers degree of LL.D. on Ramsay, 347
Glendalough (Galway), 342
Glengariff, 217
Glen Sannox, 23
Glyder fawr, 107, 136, 148, 149, 154, 255
—— fach, 150, 151
Goatfell, view from top of, 22
Godwin-Austen, R. A. C., 78, 124, 144, 309, 334
Goodsir, J., 222
Gorner glacier, 267
Graham, J., notice of, 29; at University College, 118
Grant, Dr., 118
Grantham, 305
Green, A. H., 309

Peach, B. N. (Geological Survey), 294, 342
Peaks, Passes, and Glaciers, 199, 260
Peel, Sir R., sounded as to becoming President of the Royal Society, 128; on Jermyn Street Museum, 143; at Geological Society, 144
Pembrokeshire, geology of, 43, 50, 231, 232, 233, 234
Pentland, J. B., 146
—— Hills, 320
Pen-y-bont, 81
Pen-y-gwryd, 150, 154
Percy, Dr. John, 145, 186
Permian breccias, investigation of, 198, 219, 228, 229, 241, 362
Perth, 344
Petrography in the Geological Survey, 343, 355
Phillips, J., notice of, 18; employed in Geological Survey, 43, 50, 63, 162
—— J. A., 129, 130, 176
—— R., Curator of Museum of Economic Geology, 41, 66; death of, 185
—— W., 58
Physical Geology and Geography, by A. C. Ramsay, 277, 311, 334, 344, 363
Physiography, Ramsay's contributions to, 358
Playfair, John, his *Illustrations of Huttonian Theory*, 116
Playfair, Lord, an early friend of Ramsay, 9; his reminiscences of Sir Charles and Lady Lyell with Ramsay in Arran, 14; with Ramsay in London, 61, 62, 64, 77, 78, 144, 176; appointed chemist to Geological Survey, 66; with the surveyors in Wales, 108; at the time of the Chartist riot, 129; appointed to School of Mines, 186
Plynlimmon, 75
Po, River, article by Ramsay on the, 323
Pont de Gard, 353
Ponteland, 303
Porteous Mob, anecdote of, 3
Portlock, J. E., 65
Port Madoc, 133, 326, 329
Portrush, 342
Pre-Cambrian rocks of Anglesey, 154, 172, 191, 192, 207
Preston, 305
Prestwich, J., 78, 124, 144, 305, 311
Privy Council, Committee on Education, 208

Professor, work of a, 117
Protichnites, reading of Logan's and Owen's papers on, 197
Pumpsaint, Ramsay stationed at, 76, 235, 331
Punch on the quarrels of the geologists, 283
Purbeck formation, Forbes' subdivisions of, 180
Puy de Dôme, 353
Pwlheli, 153

RAIN, geologising in, 46, 105, 113, 151
Ramsay, A. C.—*Chap.* I.—1814-1840.
—Parentage, 1; brothers and sister, 3; birth, 4; childhood, 4; attends Glasgow Grammar School, 5; first reads Shakespeare, 6; loses his father, 7 (*see* p. 257); goes into business, 7, 257; edits a manuscript journal, 8; enters into partnership as a cloth-merchant, 9; influenced towards science by D. Landsborough, 13; and by J. P. Nichol, 14; makes the acquaintance of Lyell, 15; geologically surveys Arran, 17; first meets De la Beche, Murchison, E. Forbes, etc., 18; reads his first geological paper, 19; misses the excursion to Arran, 20; writes his volume on Arran, 21; extracts from this volume, 22-25; leaves home to join Murchison, 26; letter to his mother, 26; to his brother William, 27; first visits London, 28; is appointed to the Geological Survey, 28; dines at Geological Club, 30; joins the Survey at Tenby, 33 (*see* pp. 176, 234)
Chap. II. — 1841-1845. — His opinion of the early maps of the Geological Survey, 39, 53; early days in the Survey, 42, 49, 234; with J. Phillips in South Wales, 43, 50; first Report on Welsh Geology, 51; first mapping of volcanic rocks, 51; sings in cathedral choir at St. David's, 54, 234; musical talent of, 54; sociality of, in South Wales, 55; finds fossils in South Wales, 56; undertakes, with E. Forbes, to prepare a new edition of Conybeare and Phillips' *Geology*, 58, 72; life in London, 61; promoted to be Local Director, 61; writes the essay on the "Denudation of South Wales," 63; declines the

THE END

Printed by R. & R. CLARK, *Edinburgh.*

Printed in the United States
By Bookmasters